D1498664

DNA COMPLEX
AND
ADAPTIVE BEHAVIOR

JOHN GAITO

York University

PRENTICE-HALL, INC., Englewood Cliffs, New Jersey

To F. H. C. Crick,
whose *Scientific American* articles of 1954 and 1957
suggested to me the possibility
of integrating Psychology and Molecular
Biology as Molecular Psychobiology

Prentice-Hall Series in Experimental Psychology
JAMES J. JENKINS, Editor

CONTENTS

PREFACE

This book attempts to discuss some of the interesting research, theories, and problems of the past decade within Molecular Psychobiology and related areas, especially Molecular Biology, and suggests the possibility of developing a type of chemotherapy called DNA Complex Therapy. The orientation throughout is a molecular biological one. The complex behavior of an organism has its foundations in intra- and intercellular events which rely on gene and gene product interactions. My attitude is that an efficient approach to understanding behavior is to extend the ideas and techniques of the molecular biologist to the study of adaptive behavior in higher organisms.

This work is intended mainly for individuals with some knowledge of molecular biology and psychology; however, definitions and descriptions are provided liberally throughout the text and in a Glossary for those with lesser backgrounds. For convenience the terms in the Glossary are divided into two parts: Biological-Chemical and Psychological-Behavioral.

Acknowledgements

I have been influenced in my thinking by interaction with many individuals, especially graduate students, too numerous to mention. I am especially grateful to M. Miyagi and J. Burtis for reading a preliminary draft and offering comments.

I want to thank the following for permission to use the materials indicated: J. Bonner and the Editors of *Science* for Table 2, from Bonner et al. (1968); E. Harbers and G. T. Varlag for Tables 3 and 5, from Harbers, Domagk, and Muller (1968); U. Clever and Holden-Day, Inc. for Table 4, from Clever (1964); H. Hydén and the Editors of the *Proceedings of the National Academy of Sciences* for Tables 6, 7, 8, 9, 10, and 11, from Hydén and Egyhazi (1962, 1963, 1964); the *Journal of Neurochemistry* and Pergamon Press for Table 12, from Hydén et al. (1969); G. Zubay and Holden-Day, Inc. for Figure 6, from Zubay (1964); and B. W. Agranoff and the Editors of *Science* for Table 14, from Agranoff and Klinger (1964).

Preparation of this manuscript was facilitated by grants from the National Research Council (Canada) and by Contract No. Nonr 4935(00) with the Physiological Psychology Division of the Office of Naval Research (U.S.A.).

Chapter 1

INTRODUCTION

In recent years tremendous advances have occurred in an area of biology (Molecular Biology) which has emphasized the molecular aspects of biological organisms. Molecular Biology has been concerned mainly with the structure and function of the nucleic acids, deoxyribonucleic acid (DNA) and ribonucleic acid (RNA). DNA has been established as genetic material. The role of the nucleic acids in translating genetic information into the synthesis of specific proteins has been analyzed thoroughly. RNA codes have been proposed for each amino acid. This rapid progress has generated an air of excitement and enthusiasm and filled researchers with the belief that man is getting closer and closer to nature's basic secrets.

In that the linear sequence of bases in DNA provides the information which specifies the genetic potentialities of an organism, it was immediately evident to a number of behavioral scientists that the nucleic acids might function directly or indirectly in behavioral events. Thus an area of research (Molecular Psychobiology) has developed which is attempting to integrate the ideas and methods of psychology and related behavioral sciences with those from molecular biology. At the moment a number of interdisciplinary research teams are concerned with the possible role that the nucleic acids and proteins might play in learning, sensory and motor events, emotional stress, and other behaviors. The main interest of most researchers has been in learning phenomena.

The emphasis of these individuals has tended to be on the nucleic acids, DNA and RNA, and related macromolecules. It should be obvious, however, that the nucleic acids perform their function through the involvement of a sequence of neurochemical events. Thus this approach has focused on a narrow segment of the problem. For example, in a specific learning situation a sequence of events is involved, say, $a, b, c, \ldots, x$. Event a could be stimulation of receptors, b the transmission of nerve impulses into the central nervous system, etc. Another event, say *g*, would involve DNA; *h*, RNA synthesis; and *i*, protein synthesis. The molecular psychobiological approach has been concerned with events *g*, *h*, and *i* of the overall sequence. What occurs before *g* and after *i* is not specified. However, an understanding of certain aspects of *g*, *h*, and *i* should help to suggest the antecedent and postcedent conditions prevailing. Furthermore, these events are of such basic nature that information obtained at this level should be extremely valuable for biological science.

The molecular biological contributions which are most pertinent are those concerned with heredity and the manner in which DNA performs its genetic function. The importance of molecular events in genetics and behavior cannot be overemphasized. For example, the difference between a normal individual and some anemics (sickle cell anemia) is due to one amino acid in hemoglobin. In sickle cell anemia 30 to 60 percent of the red blood cells (erythrocytes) form a sickle shaped structure in contrast to normals whose red cells do not sickle (Strauss, 1960). The sickle cell erythrocytes contain a hemoglobin in which one amino acid (valine) is substituted for another (glutamic acid) at one site in the protein chain. A number of other abnormal hemoglobins have been reported which depend on an amino acid change at the same site (Strauss, 1960). The genetics of these conditions are relatively simple, resulting from a single recessive gene. Likewise, slight modification in one amino acid in insulin may vitiate its usefulness for diabetic persons. The implication of these results is that the DNA and RNA for the normal condition differs slightly from that of the abnormal condition.

Another example of the drastic effect which a simple genetic defect can have on a human is the condition called phenylketonuria. The symptoms include a retardation in intelligence, a possibility of epileptic seizures, and occasionally psychotic episodes. The disorder is identified by an accumulation and excretion in the urine of phenylalanine (an amino acid) and its pyruvic, lactic, acetic, and other derivatives. Within a few weeks after birth, phenylpyruvic acid has inflicted permanent damage in brain tissue. Fortunately, the defect can be detected promptly, and if the children are placed on special diets low in phenylaline, they can escape nearly all the brain damage.

This disorder is due to a block in the transformation of phenylalanine to another amino acid, tyrosine. A protein fraction in the liver required as an enzyme in this transformation is absent. Thus a block at a single point in a series of metabolic events leads to disastrous effects.

The genetics of phenylketonuria are relatively simple; this condition is inherited as though it were controlled by a single recessive gene. In this case a protein which functions as an enzyme is absent because the appropriate genetic code for the protein does not appear in the DNA composition.

As our understanding of molecular events grows, it should be possible to alter human and lower organisms through chemical means. It is already possible to induce a permanent change in some viruses and bacteria which is passed on to later generations. It may be possible *to create artificially* relatively harmless viruses which would crowd out disease producing ones and to manufacture a virus which would destroy cancerous cells but ignore healthy ones.

Microbiologists at the University of Illinois (Mills, Peterson, and Spiegelman, 1967) have reported that a harmless virus RNA particle has been developed. This artificial virus particle (devoid of portions which make it harmful) competed favorably with a natural virus and prevented the latter from reproducing itself and infecting other cells. This event resulted in an exclusion of the viral infection. The artificial virus particle was created by taking RNA from a natural virus and using it as a template to make more RNA; then this latter RNA was used as a template to make RNA again. After repeating this process 74 times, the RNA was quite different from the original RNA template. The RNA strands were much shorter, and only 17 percent of the original gene remained; the genetic information encoded in certain sequences of the RNA to make the virus protein coat was lost. A most important difference was that this artificial RNA would reproduce itself 15 times as rapidly as would the natural virus by taking for itself all the host cell materials which are required for viral replication. Artificial DNA virus particles have been developed also (Kornberg, 1968).

One of the most striking possibilities which may be achieved soon by mankind is that of controlling his own evolution. In the past, evolutionary trends have been determined by natural forces. Now that the mysteries of DNA and RNA are being unravelled and DNA and RNA capable of reproducing themselves have been synthesized in the test tube, the ability to provide human organisms with certain desired characteristics may follow soon. If scientists are able to discover the genetic code for various human potentialities, synthesis of DNA with the code appropriate for certain "desirable" characteristics might be achieved. For example, man might be able to modify his intelligence, his emotional well-being, his facial and body features, his susceptibility to disease, his life span, and many other characteristics by DNA control. This DNA modification might occur at conception or at later times.

The control of DNA is only one means for possible control of human characteristics, although it is the most basic one. The same results might be precipitated by dealing with the products of DNA function, i.e., RNA and protein molecules. For example, RNA therapy has been reported by some investigators to improve the overall condition of elderly patients. Other chemicals such as magnesium pemoline and strychnine sulfate which allegedly improve the performance of organisms may operate to influence DNA or other molecules or processes related to DNA such as RNA and protein synthesis. Thus it should be obvious to the reader that a type of therapy which one might designate as *DNA complex therapy* may achieve greater significance in the future than electric shock therapy, chemical therapy, surgical therapy, and other therapies in the behavioral sciences have in the past. The aim of this book is to describe the DNA complex, discuss research results related to this complex, and to suggest possible means of using DNA complex therapy to improve the performance of humans in educational and other applied situations.

Chapter 2

PRIMARY, SECONDARY, AND TERTIARY
STRUCTURES IN DNA COMPLEX

The term, *DNA complex,* is used in this book to designate DNA and the molecular structures which appear to be intimately involved in its function, specifically RNA and protein.

DNA is a large double stranded molecule which is found in the nucleus of cells (Figure 1). It is helical in shape (like a spiral staircase) and is about 20 angstroms (Å) in diameter, but very long—up to several millimeters. The strands of DNA in a single human cell would reach nearly six feet if stretched to full length. This amounts to some 10 billion miles of DNA in every man and woman; Stanley and Valens (1961) suggested that in humans there are 800,000 DNA molecules each with about 40,000 nucleotides, or about 32 billion nucleotides in all. Other estimates (Tatum, 1964) have been lower, however, only five billion nucleotides. In the rat an estimate of 750,000 DNA molecules, each with about 15,000 nucleotides, has been offered.

FIGURE 1 Two stranded DNA molecule in helical form.

Each DNA strand consists of a recurring pattern of constituents, called *nucleotides.* Each nucleotide contains a phosphate group attached to a sugar-base linkage. The nucleotide, deoxyadenylic acid, is shown in detail in Figure 2. The sugar-base linkage is called a *nucleoside.* The nucleoside resulting from the removal of the phosphate group in Figure 2 is deoxyadenosine.

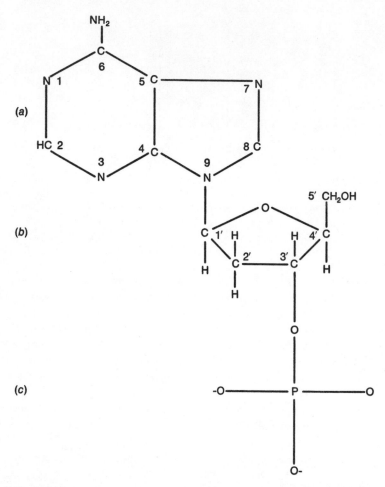

FIGURE 2 The nucleotide, deoxyadenylic acid. The numbers refer to the position in the molecule and are used for identification purposes. *a* is the purine base adenine; *b*, deoxyribose sugar; *c*, phosphate group.

Figure 3 indicates a portion of a DNA molecule as it would appear if it were unwound from its helical formation. The bases of one strand are attached to bases of the other strand by hydrogen bonds. The bases consist of two types: purines and pyrimidines. There are two purines, adenine (A) and guanine (G), and two pyrimidines, thymine (T) and cytosine (C). The purines are larger molecules than are the pyrimidines. A purine of one strand is always attached to a pyrimidine since two purines are too big to bridge the gap between the two strands and two pyrimidines are too small. Furthermore, the amounts of A and T are equal as are the amounts of G and C; A is paired with T and G with C as shown in Figure 3. The binding

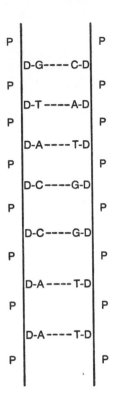

FIGURE 3 A portion of the DNA molecule as it would appear if unwound from its helical formation. P is phosphate; D, deoxyribose; A, adenine; T, thymine; C, cytosine; and G, guanine.

of T to A and G to C indicates the complementarity of the two strands. These two pairings are shown in detail in Figure 4. The triple hydrogen bonding between G and C is stronger than the double bonding of A and T. The ratio of these two types of pairs $(A + T/G + C)$ varies from .51 in *Pseudomonas aeroginos* (a bacterium) to 1.86 in the sea urchin. In trout, domestic chicken, horse, and man, this ratio is 1.3 to 1.4.

The sequence of the bases along a DNA strand must furnish the "code" of genetic potential inasmuch as DNA carries genetic information and the main difference between DNA molecules appears to be in the sequence of these bases. Even though there are only four different nucleotides, the strands of the molecule are extremely long and allow for many different sequences of these nucleotides. DNA molecules consist of thousands of nucleotides, and even assuming that only 1000 units are present in a single DNA molecule (a figure which is much less than that which occurs even in the smallest virus), the number of different sequences possible would be 4^{1000}. This figure is tremendous. Further, when we consider that there are many

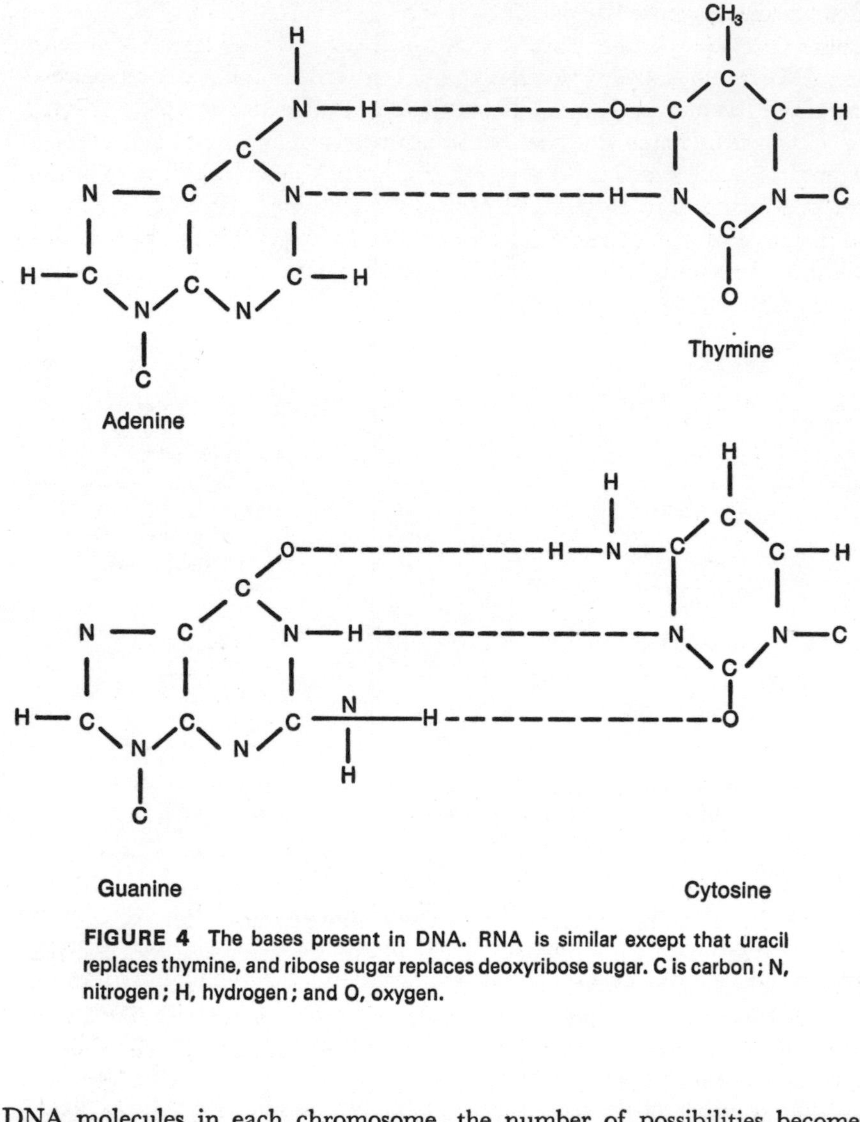

FIGURE 4 The bases present in DNA. RNA is similar except that uracil replaces thymine, and ribose sugar replaces deoxyribose sugar. C is carbon; N, nitrogen; H, hydrogen; and O, oxygen.

DNA molecules in each chromosome, the number of possibilities becomes incomprehensible. Thus it is obvious why biological scientists believe that the sequence of bases could furnish the "language" of the genes. Crick (1954, 1957) stated that there is enough DNA in a single cell of the human body to encode about 1000 large textbooks.

Watson (1965) estimated that an average gene consists of 1500 nucleotide pairs. Thus the possible number of genes using the four letter code of DNA would be 4^{1500}, a value much larger than the number of different genes that have existed in all the chromosomes since the origin of life.

RNA is found in the nucleus and in the cytoplasm (approximately 10 and 90 percent, respectively). RNA is similar to DNA in most aspects but contains uracil (U) rather than T and ribose sugar rather than deoxyribose sugar. Ribose has a hydroxyl group (OH) at the second carbon, whereas deoxyribose has only an H atom there (Figure 2); thus deoxyribose is missing an oxygen, as the name indicates. Another important difference is that RNA is a single stranded molecule, although some RNA's contain loops in which bases bond to nearby bases to provide double strand molecular structure. The diameter of RNA molecules is 10 to 15 Å; their length is up to several thousand angstroms. The names of the bases, nucleosides, and nucleotides of RNA and DNA are given in Table 1.

TABLE 1   Bases, Nucleosides, and Nucleotides of The Nucleic Acids

	Base	*Nucleoside*	*Nucleotide*
DNA	adenine	deoxyadenosine	deoxyadenylic acid
	guanine	deoxyguanosine	deoxyguanylic acid
	cytosine	deoxycytidine	deoxycytidylic acid
	thymine	thymidine	thymidylic acid
RNA	adenine	adenosine	adenylic acid
	guanine	guanosine	guanylic acid
	cytosine	cytidine	cytidylic acid
	uracil	uridine	uridylic acid

There are three basic types of RNA, all of which are synthesized by DNA in the nucleus and then proceed to the cytoplasm: messenger RNA (mRNA), transfer or soluble RNA (tRNA), and ribosomal RNA (rRNA). RNA particles in the nucleolus (a nuclear substructure) bear at least a superficial resemblance to rRNA (Swift, 1962). Chipchase and Birnstiel (1963) showed that nucleolar RNA in plants was identical to rRNA in base sequence and that the DNA regions complementary to rRNA were not confined to the nucleolus but were scattered throughout the nucleus, comprising approximately 0.30 percent of the total DNA sites. This percentage has been found also with bacteria (Yankofsky and Spiegelman, 1963) and with an insect, Drosophilia (Ritossa and Spiegelman, 1965). Perry, Srinivasan, and Kelley (1964) found also that nucleolar RNA appears to be the precursor of rRNA. Using radioactive labeling, Comb, Brown, and Katz (1964) found that the nucleolus was the site of rRNA synthesis whereas the chromatin (chromosomes in extended form between mitotic cell divisions) was the site for mRNA synthesis.

Jacob and Monod (1961) introduced the concept of mRNA. The characteristics which were included in the definition of this type of RNA are the following (Jacob and Monod, 1961; Grunberg-Manago, 1963):

1. A base composition (i.e., the proportion of different kinds of bases) reflecting the DNA base composition with long base sequences which are complementary to specific DNA sequences.

2. Heterogeneity in molecular weights reflecting the different sizes of the protein chains to be synthesized.

3. Association with ribosomes during protein synthesis.

4. High turnover, indicating a short life.

5. Stimulation of protein synthesis (*in vivo* and *in vitro*).

Although one might consider that (1) is the characteristic most useful in differentiating mRNA from tRNA and rRNA, this is not the case. Studies suggest that all RNA is synthesized on DNA sites; thus each RNA molecule would have to satisfy condition (1). One possible exception may be tRNA which is assumed to be synthesized in the nucleus by DNA but has the terminal three nucleotides added in the cytoplasm.

Characteristic (4) is a useful one for differentiation purposes, although some long lived mRNA's have been reported, e.g., in ocular lens and feather cells of chicks (Scott and Bell, 1964).

Characteristic (5) is the basic one for the messenger concept and is the most useful for differentiation purposes (Barondes, Dingman, and Sporn, 1962).

The notion of mRNA was developed from studies with bacteria. It has been shown, however, that this type of RNA is also present in mammalian cells (e.g., Barondes, Dingman, and Sporn, 1962; Scott and Bell, 1964; Sonneborn, 1964; Harbers, Domagk, and Muller, 1968).

Ribosomal RNA is a constituent of cellular structures called ribosomes. Ribosomes are approximately 100 to 200 Å in diameter and consist of about half protein and the other half RNA; in general the RNA varies from 30 to 65 percent depending on the species involved (Grunberg-Manago, 1963). For example, ribosomes in goat brain cortex have been found to contain 30 percent RNA and 70 percent protein and differ in RNA-protein proportions from similar preparations from the liver (Datta and Ghosh, 1964). The ribosomal structure is strongly dependent on the amount of magnesium present in the medium. If magnesium is not present, the large particle breaks down into smaller ones. In bacteria the ribosomes probably exist as 70 S (Svedberg constant, based on rate of sedimentation) units which can dissociate to 50 S and 30 S particles (Loewy and Siekevitz, 1963; Watson, 1965). The 30 S particles contain one molecule of 16 S RNA and about 10 molecules of protein while 50 S particles have one 23 S unit and 20 molecules of protein, with the RNA appearing to be on the surface of the ribosome (Santer, 1962). In mammalian cells the ribosomes are somewhat bigger and more stable than those from bacteria; they consist of 80 S particles. These can be split to 60 S and 40 S units, which contain one molecule of 29 S and 18 S RNA, respectively (Harbers, Domagk, and Muller, 1968). The proteins of the ribosomes seem to be somewhat similar to, but not the same as, the histones which are found in the nucleus. The base composition of the 30 S, 50 S, and 70 S RNA particles in *Escherichia coli* (*E. coli*) is roughly the same (Grunberg-Manago, 1963).

Recently another type of ribosomal RNA (5 S) has been reported (Fellner and Sanger, 1968). The 5 S RNA comprises about 2.5 percent of the total content of *E. coli* ribosomes. The nucleotide sequence of this RNA has been determined; portions of the sequence of the 23 S and 16 S RNA's are known also. These latter RNA's contain small amounts of methylated nucleotides. Fellner and Sanger (1968) were able to investigate the sequence in the neighborhood of the methylated sites; the sequences occurred twice, as if each molecule consisted of two identical molecules or as if each were one chain with a duplicated base sequence. This partial duplication has been found in 5 S RNA from *E. coli* also. These results suggest that the ribosomal RNA's are homogeneous, at least in the methylated region.

X-ray diffraction studies of *E. coli* ribosomes have indicated that the substructure consists of linear aggregates of particles and that the RNA and protein components are somewhat independent (Langridge and Gomatos, 1963). Detailed analyses suggested that an array of four or five parallel RNA double helices, 45 to 50 Å apart, were involved; the protein component might fulfill a structural function in preserving the spacing in the parallel RNA, perhaps interacting with the RNA in the nonhelical regions. The 45 to 50 Å spacing was common to ribosomes from *E. coli, Drosophilia* larvae, rat liver, and rat reticulocyte.

Transfer RNA is the smallest of the nucleic acids, containing only 50 to 100 base units with a length of approximately 250 Å when unfolded. There are a number of different tRNA's; each one appears to be a single strand looped molecule with profuse base pairing in the loops. Each tRNA is assumed to attach to a specific amino acid and proceed to a ribosome during protein synthesis. Each terminates in the sequence CCA at one end and G at the other end. There is evidence that in most samples the fourth nucleotide, next to cytidylic acid, is either an A or G nucleotide (Smellie, 1963). The fourth nucleotide contains A in 69 percent of *E. coli* RNA and in most of the rat liver RNA (Grunberg-Manago, 1963). The trinucleotide portion which is specific for mRNA pairing (see Chapter 4) is presumed to be located in the central part of the molecule (Nirenberg, 1963; Wilkins, 1963). Zubay (1963) utilized the available physical-chemical data to develop a hairpin-like molecular model of tRNA with 67 bases (Figure 5*a*). The folded portion with secondary helical structure consists of thirty base pairs encompassing about three turns of the helix. Of the seven remaining bases, only five are unpaired; the G at one end and the innermost C of the other end pair through a hydrogen bond; in Figure 5*a* the AAA trinucleotide unit and the terminal C and A are unpaired.

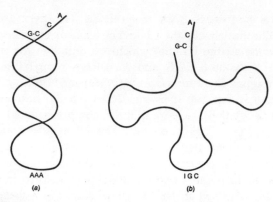

FIGURE 5 *a*. The hairpin conception of the structure of a transfer RNA. Base pairings are assumed to occur in each of the three loops. *b*. The cloverleaf model for alanine transfer RNA in simplified form. Base pairings are present in the stem of each leaf. In both models bases occur in sequence along the strand (represented by lines in the drawings).

Holley et al. (1965) were able to purify three tRNA's from yeast which were specific for the amino acids alanine, tyrosine, and valine. They established that the nucleotide sequences of each were different by careful analysis of the breakdown products following ribonuclease (RNase—the enzyme which degrades RNA) treatment. They determined the complete nucleotide sequence of the alanine tRNA which contained 77 nucleotides. Their analysis of alanine tRNA suggested that the structure was in the form of a cloverleaf. The three arms of nucleotides are folded tightly together with the fourth arm extending in the opposite direction (Figure 5*b*). The cloverleaf model for tRNA's has gained wide acceptance (Singer, 1968). The I in the IGC sequence is inosine, a nucleoside (purine base with ribose), which is found infrequently in nucleic acids.

This marvelous achievement of determining the nucleotide sequence of an RNA molecule revealed for the first time the complete primary structure of a gene (Sonneborn, 1965). As will be explained in Chapter 4, the base sequences for tRNA's are coded by certain DNA sites. Thus, knowing the complete sequence of bases in the alanine tRNA allows one to determine the DNA sequence which is responsible for this RNA.

The complete nucleotide sequences for two tRNA's for serine and one each for tyrosine and phenylalanine have since been determined (Harbers, Domagk, and Muller, 1968). The total number of nucleotides for the serine RNA's were 84; for tyrosine, 78; and for phenylalanine, 76.

It seems probable that tRNA is synthesized in the nucleus and proceeds to the cytoplasm where it has the CCA terminal sequence added because the enzymes responsible for this addition appear to be located in the cytoplasm (Smellie, 1963).

An interesting observation concerning tRNA's is that they contain a number of unusual bases: pseudouracil, thymine, 2-methyladenine, 6-methyl-aminopurine, 1-methylguanine, 2-thiocytosine, as well as others; the exact function of these bases is unknown (Srinivasan and Borek, 1964; Carbon, David, and Studier, 1968). Ribosomal RNA contains some of these odd bases also; however, since tRNA is considered to interact with ribosomes, it is possible that some of these bases originate in tRNA and are trapped in the ribosomes.

The enzymes, called RNA methylases, whose apparent function is the introduction of methyl (CH_3) groups on tRNA, are concentrated in the nucleolus. This event suggests that the methylation of tRNA occurs here (Birnstiel, Fleissner, and Borek, 1963), although other evidence suggests that methylation occurs in the cytoplasm (Comb and Katz, 1964).

The base composition of tRNA differs from that of rRNA (Grunberg-Manago, 1963). Nuclear RNA differs from cytoplasmic RNA; in guinea pig brains, the nuclear RNA contains significantly greater amounts of U and less G than do other subcellular fractions (Yamagami, Kawakita, and Naka, 1964).

Bonner (1965) found a fourth type of RNA adhering to histones in plants (chromosomal RNA), possibly providing a specificity for the latter's complexing with DNA. The chromosomal RNA was bound in the structure of chromosomal material in such a way as to be resistant to RNase, suggesting that it was bound to DNA by base pairing. This RNA was characterized by short chain length (40 to 60 nucleotides), a sedimentation coefficient of 3.2 S, and a relatively high content (5 to 25 percent) of dihydrouridylic acid (Bonner et al., 1968).

Another RNA, activator RNA, has been suggested by Britten and Davidson (1969). This RNA is confined to the nucleus and, in the regulation model of Britten and Davidson, serves to turn on protein synthesis (see Chapter 4).

Although the existence of the chromosomal RNA of Bonner and the activator RNA of Britten and Davidson are suggested by research results, it is possible that both are merely special classes of mRNA.

Using ingenious procedures, molecular biologists have been able to obtain estimates of the percentage of total DNA which is involved in the synthesis of various RNA species. Even though rRNA comprises up to 80 to 90 percent of the total RNA in the cell, Yankofsky and Spiegelman (1963), Chipchase and Birnstiel (1963), and Ritossa and Spiegelman (1965) indicated that about 0.30 percent of total DNA sites are concerned with its synthesis in plants, bacteria, and insects. Likewise, Goodman and Rich (1962) found that 0.025 percent of DNA was responsible for the synthesis of tRNA. In rat brain tissue, Stevenin et al. (1968) estimated that 0.15 percent of DNA is responsible for rRNA. Thus the synthesis of both these RNA's is determined by less than one percent of the total DNA. The synthesis of mRNA would involve larger portions of the DNA sites. Stevenin et al. (1968) reported that the percentage of DNA sites coding for mRNA in adult rat brain was 1.2 or greater. Probably a small percentage of DNA sites would code for mRNA's in each of the various tissues of an organism, e.g., liver, kidney, muscle. However, some DNA sites would appear to have a function other than coding for these RNA species, because it is assumed that DNA's operate for events other than protein synthesis, e.g., in the synthesis of activator RNA (Britten and Davidson, 1969) to control other DNA sites. Current thought in genetics is that there are three or four types of genes (or DNA), some involved in the direct synthesis of RNA's for protein synthesis and others regulating these genes (Harbers, Domagk, and Muller, 1968; Britten and Davidson, 1969).

The above description concerns the primary and secondary structure of the nucleic acids, i.e., the linear sequence of bases (primary) and the H bonding across helices (secondary).

Secondary structure is expressed in DNA by the hydrogen bonding between bases on the two strands. Some RNA's, which are of single strand nature, have secondary structure with profuse bonding presumed to occur throughout the helical portions. Both tRNA and rRNA appear to have some secondary structure; mRNA does not, however; secondary structure appears to interfere with its biological activity (Nirenberg, 1963).

Tertiary structure is concerned with the overall configuration of the molecules; the nucleic acids can exist in a number of configurations. For example, DNA's vary in the number of base pairs per turn of helix and in the inclination of base pairs in the molecule (Wilkins, 1963). Both nucleic acids vary their configurations somewhat with some changes in the chemical environment. DNA is a relatively rigid rod in its natural state. Upon denaturation, both nucleic acid chains form random coils upon the breaking of hydrogen bonds.

The above description indicates the usual state of affairs. There are, however, a number of exceptions which should be considered.

1. In the DNA of a few bacteriophages, e.g., T2, T4, and T6, glucose (a six carbon or hexose sugar) is found in addition to deoxyribose (a five carbon or pentose sugar).

2. The DNA in T2, T4, and T6 (bacteriophages which infest *E. coli*) have 5-hydroxymethyl-cytosine in place of C. The former contains both OH and CH_3 groups at the carbon 5 position.

3. Although not a normal base, U has been reported in DNA (Belozersky and Spirin, 1960).

4. Likewise, T has been found in RNA, e.g., in tRNA, in wheat germ, in *E. coli,* and in other bacteria (Belozersky and Spirin, 1960).

5. The information contained in the second strand of DNA must not be essential in terms of information content because *in vivo* studies show that RNA is synthesized from only one of the two strands of DNA.

6. Most organisms contain DNA which performs the genetic function. Some viruses, however, contain only RNA as the genetic material. Those with RNA include polio, influenza, and tobacco mosaic viruses.

7. Some viruses have double stranded RNA. For example, reoviruses which inhabit the respiratory and enteric tracts of man and animals have a highly ordered double stranded helix (Gomatos and Tamm, 1963).

8. Extranuclear DNA has been reported in mitochondria of both plants and animals and in the plastids of plant cells (Gibor and Granick, 1964; Granick, 1965). At first it was assumed that the DNA found in these structures were nuclear contaminants; however, more careful preparations have shown that these particles have their own DNA (Harbers, Domagk, and Muller, 1968) and that the mitochondria are able to replicate their DNA (Karol and Simpson, 1968). These structures also contain RNA; in mitochondria messenger, transfer, and ribosomal RNA's are present (Harbers, Domagk, and Muller, 1968).

Chapter 3

QUATERNARY STRUCTURES
IN DNA COMPLEX

Primary, secondary, and tertiary structures relate to the characteristics of single molecules. Quaternary structure refers to the interaction or complexing of a nucleic acid with other nucleic acids or proteins or other molecules, specifically a DNA complex. There are two types of quaternary structures which are of significance for this book. One is concerned with enzyme induction and repression phenomena and the other, with a specific DNA-protein interaction.

It has been indicated that the presence of certain chemicals within an organism can later lead to an increase in the amounts of specific enzymes or other proteins available, or to a decrease. Examples are enzyme induction and repression phenomena, and the effects of hormones on DNA activity. One specific mechanism which has been suggested for these events is that a regulatory gene (DNA) codes for a repressor substance consisting probably of a protein (Jacob and Monod, 1961; Watson, 1965). If a corepressor (possibly excessive amounts of a chemical) complexes with the repressor substance, the complex interacts with other DNA (operator gene sites). This process inhibits the functioning of structural genes in making mRNA for the synthesis of a specific protein. The overall result is complete cessation of, or a slow down in, production of the specific protein. If the repressor substance complexes with another chemical, an inducer, the interaction with the operator gene is prevented and leads to an increased production of the protein. Another mechanism suggested by Britten and Davidson (1969) is that the hormone or other inductive chemicals affect specific DNA sites such that an activator RNA is synthesized which controls other DNA sites. Results which suggest regulation at the gene level have been reported (Hayashi et al., 1963; Harbers, Domagk, and Muller, 1968).

Hurwitz and August (1963) suggested that different species of tRNA may act as repressors of RNA synthesis, and amino acids, as inducers. Thus each specific repressor tRNA would be "neutralized" by a specific amino acid in the formation of an RNA-amino acid compound.

Pardee (1962) suggested that genes or ribosomes could be the site of attachment for the repressor substance. He indicated that the ribosome was the most popular candidate because of the ease in visualizing the repressor as blocking the laying down of amino acids on the template or preventing the release of the finished enzyme. In line with this suggestion are the results of Hoagland, Scornik, and Pfefferkorn (1964). They reported inhibition of protein synthesis by a component which appeared to be in the membranous portion of the microsome. The inhibition was reversed by guanosine triphosphate (GTP); there was a direct proportionality between inhibitor concentration and degree of GTP stimulation. They suggested that GTP played a role in normal regulatory processes. Ohtaka and Spiegelman (1963) also reported results showing regulation at the point of protein synthesis.

A second quaternary structure, and a potential regulatory mechanism, involves the complexing of DNA with proteins in multicellular organisms. Proteins consist of a number of amino acid molecules joined together in a long chain, with side branchings. For example, insulin has 51 amino acid units. There are 20 common amino acids in nature, all of which contain one or more amino (NH_2) and carboxyl (COOH) groups. Two amino acids link together through the amino and carboxyl groups, with the release of one molecule of water (H from NH_2 and OH from COOH).

The function of proteins in chromosomal material is of great concern; the chromosome involves a complexing and interaction of DNA, RNA, and proteins. Much of the RNA that has been reported in chromosomes may be newly synthesized strands but some RNA's may perform supportive or regulatory roles. The proteins presumably serve some supportive role also, although some proteins may serve a regulatory function.

Chromosomes have been studied usually in their extended form between mitotic cell divisions because in this condition they carry out both DNA replication and RNA synthesis. In this condition the chromosomes are called *chromatin*. Such chromatin is composed of DNA complexed with proteins and some RNA and RNA polymerase (the enzyme required in RNA synthesis); only a portion of the genetic material is available for RNA synthesis.

In the chromosomes of higher organisms, there is approximately 50 percent protein. There are two types of proteins in the chromosomes: non-histones and histones. The latter have received more attention because of their apparent inhibitory effect on RNA synthesis. The histones are basic (have a net positive charge), and it is believed that they neutralize part of the negative charge of DNA molecules (Watson, 1965). Histones are not present in bacteria. Stedman and Stedman (1950) had assigned an important function to these proteins, that of controlling gene activity. A number of other individuals (e.g., Bonner, see Chapter 5) have suggested the same. Histones have a high content of basic amino acids (mainly arginine and lysine). They are a heterogeneous group of proteins; at present 12 to 20 histones are believed to be present in varying amounts in cells; however, the histones have been fractionated into four components by certain methods. These fractions have been denoted as F1 (very lysine rich), F2a (lysine rich), F2b (lysine rich) and F3 (arginine rich). In calf thymus F1 contains 30 percent lysine and 22 percent alanine; F2a—11 percent lysine, 12 percent arginine, 13 percent glycine, 10 percent alanine, and 10 percent leucine; F2b—16 percent lysine, 8 percent arginine, and 11 percent alanine; and F3—10 percent lysine, 13 percent arginine, and 13 percent alanine. In each fraction there were small amounts of other amino acids (Billen and Hnilica, 1964). The histones tend to be intimately associated with DNA; that they function intimately with DNA is suggested by their close association and by the fact that the amounts of DNA and histones vary together. For example, in the RNA-rich chromosomal puffs in certain insects, the amounts of DNA and histones remain constant (Swift, 1962). In the DNA-rich puffs of the insect *Sciara* both histones and DNA increase in amount. Thus in both RNA and DNA puffs, histones follow the pattern of DNA.

Nonhistone proteins in the nucleus are of acidic nature. Their importance metabolically is evident from their high rates of turnover, and they tend to be associated with gene activity (Busch et al., 1963). The acidic proteins and histones are approximately equal in amount, each comprising 20–25 percent of the total dry weight of the nucleus in the rat liver. Busch et al. (1963) suggested that DNA and the acidic proteins compete for linkage with histones and that loss of histones from linkage with DNA to the acidic proteins would free DNA for synthesis functions.

Dingman and Sporn (1964) reported that inactive genetic material (mature red blood cells) in chickens contained negligible amounts of RNA and nonhistone proteins. RNA and nonhistone proteins were abundant in genetic material showing cellular activity: whole embryo, brain of embryo and adult, liver of embryo and adult, and red blood cell of the embryo. The ratios, nonhistone protein/DNA and RNA/DNA, varied inversely with age and were related to each other. However, the histone/DNA ratio was constant over age. The amounts of RNA and nonhistone protein declined much less markedly in brain chromatin than in the liver and red blood cells. This aspect suggested that the brain continues to use a greater variety of its genes than does the liver and red blood cells.

An acidic protein (called S 100) which is found only in the brain has been isolated recently and consists of three components (Hydén, 1967). Within 30 minutes of injection two of the three components showed high incorporation of a radioactive precursor, which suggested rapid turnover. The third showed low incorporation. The S 100 protein fraction was present in greater concentrations in glial cells than in neurons.

TABLE 2 Chemical Composition of Various Chromatins

Source of Chromatins	Content, relative to DNA, of				Template Activity (% of DNA)
	DNA	Histone	Nonhistone protein	RNA	
Pea embryo	1.00	1.03	0.29	0.26	12
Pea vegetative bud	1.00	1.30	.10	.11	6
Pea growing cotyledon	1.00	0.76	.36	.13	32
Rat liver	1.00	1.00	0.67	.043	20
Rat ascites tumor	1.00	1.16	1.00	.13	10
Human He La cells	1.00	1.02	0.71	.09	10
Cow thymus	1.00	1.14	.33	.007	15
Sea urchin blastula	1.00	1.04	0.48	.039	10
Sea urchin pluteus	1.00	0.86	1.04	.078	20

After Bonner et al., The biology of isolated chromatin. *Science*, 1968, *159*, 47–56. Copyright 1968 by the American Association for the Advancement of Science.

Table 2 indicates the relative content of DNA, RNA, and proteins in various chromatins (Bonner et al., 1968). The amounts of DNA and histone tend to be similar; the ratio of histone to DNA varies about one, from 0.76 to 1.30. For all DNA sites to be completely complexed with histones requires a ratio of 1.35 (Bonner et al., 1968).

The exact form that DNA, RNA, and histones and nonhistones take in chromosomal material is not certain. It has been suggested that at least half of the histones bound to DNA are in helical formation and that DNA-histone extends the length of the chromosomal fiber. Some investigators have proposed that the histones are wound about the DNA double helix, lying in the large groove (Sager and Ryan, 1961). Zubay (1964) suggested that histones form bridges between adjacent DNA molecules with the long axis of the helical histones at an angle of 60° to the long axis of the DNA molecules (Figure 6). He indicated that the bridges facilitate the super-coiling of DNA in a single chromosome. Before becoming active the supercoiled regions would unwind and the histone bridges would be broken and possibly reformed between less highly coiled filaments of DNA.

Based on research with giant chromosomes isolated from immature eggs from the newt, *Triturus viridescens,* Allfrey and Mirsky (1963) proposed an interesting hypothesis of the DNA-histone chromosomal structure. Each chromosome consists of long, DNA containing strands, tightly coiled or condensed in some regions (called chromomeres), and loosely extended at numerous sites along the chromosome to form "loops." The loops project outward from the main axis of the chromosome and contain DNA. The loops are sites of intensive RNA synthesis. Allfrey and Mirsky (1963) proposed the hypothesis that histones combine with DNA to produce condensed sites along the chromosome which are not active in RNA synthesis.

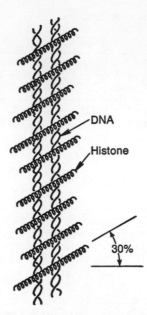

FIGURE 6 A model of the histone-DNA relationship (after Zubay, 1964).

Electron micrographs show a diameter of 20 Å for DNA whereas chromosomes have been observed in electron micrographs to be 100 to 200 Å in diameter. Ris (1964) reported that chromosomal threads in erythrocyte nuclei of the salamander were about 200 Å thick, probably representing two 100 Å fibrils. He thought that four parallel DNA strands and the histones associated with them comprised the 200 Å thread. He found similar threads in somatic nuclei of other tissues and also in plant nuclei, suggesting a general unit of organization of DNA-histones in chromosomes.

On the other hand, some individuals suggest that a linear array of single DNA molecules, rather than multiple single strands, are in the chromosome (Painter, 1964). The Zubay model of DNA-histone bridging would fit in this category.

Chapter 4

DNA COMPLEX FUNCTIONING

Protein Synthesis

The interaction of DNA, RNA, and amino acids in protein synthesis has been described frequently (Hurwitz and Furth, 1962; Ochoa, 1962; Rich, 1962; Harbers, Domagk, and Muller, 1968). The basic information (genetic code) in DNA is transcribed in mRNA in the nucleus in the presence of an enzyme, RNA polymerase. The exact manner in which mRNA is manufactured is not completely clear. However, it appears that as the two stranded DNA molecule divides, one of these strands then forms a hybrid two stranded molecule with mRNA which forms as a complement of the DNA strand. Thus if the one strand of DNA has the linear sequence ATTGC..., mRNA would consist of UAACG.... *In vitro* studies indicate that each DNA strand synthesizes a mRNA whereas *in vivo* experiments suggest that only one strand is copied (Hurwitz and August, 1963; Hayashi, Hayashi, and Spiegelman, 1963; Watson, 1965).

The newly synthesized mRNA is then transferred from the nucleus to the ribosomes in the cytoplasm where it supervises the uniting of amino acids to form proteins (Figure 7).* The transfer of RNA from nucleus to cytoplasm has been reported by numerous investigators. In the synthesis of protein, tRNA gathers an amino acid and attaches itself to its appropriate site on mRNA and on the ribosome. There are supposed to be at least two tRNA's for each amino acid. The site on the mRNA which provides for specificity is called the *codon*. The corresponding site on tRNA is the *anticodon*.

A ribosome becomes attached to one end of the mRNA molecules.

---

* Some synthesis of proteins occurs also in the nucleus (Byrne et al., 1964; Birnstiel and Flamm, 1964; Reid and Cole, 1964). Also in some cases mRNA molecules may attach to ribosomal particles prior to leaving the nucleus.

21

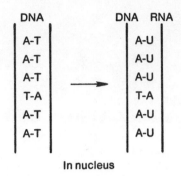

In nucleus

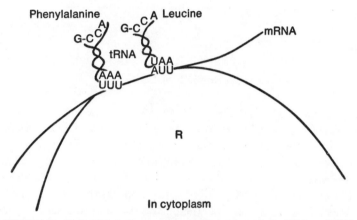

In cytoplasm

FIGURE 7 A simplified diagram showing a portion of the DNA and RNA molecules involved in protein synthesis. mRNA, messenger RNA; tRNA, transfer RNA. Only one of the DNA strands is shown as RNA is transcribed in the nucleus. The hairpin model of tRNA is used.

A ribosome is large enough to accommodate about 90 nucleotides on the mRNA at one time. As the ribosome travels along the length of the mRNA, tRNA molecules with associated amino acids bind to the appropriate site on mRNA. During this movement amino acids are joined to form a polypeptide chain and the tRNA is released. When each ribosome reaches the end of mRNA, the polypeptide chain is complete and is released from the ribosome (Figure 8). In this process a number of ribosomes are attached to a single mRNA, forming a polyribosome or polysome. The presence of polysomes in tissues suggests the occurrence of protein synthesis. In this process it has been suggested by Warner, Rich, and Hall (1962), based on electron microscopy, that a number of polysomes roll along a single messenger RNA like a ball, "reading" the message and attaching the amino acids in a polypeptide chain. Other experimental data also tend to favor this suggestion (e.g., Hardesty, Miller, and Schweet, 1963).

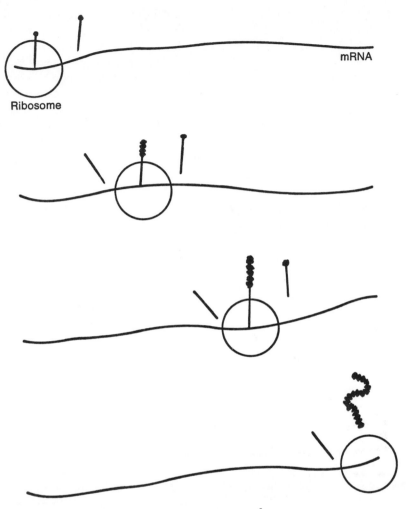

FIGURE 8 Diagram showing protein synthesis. | represents tRNA; is an amino acid; the (is the growing polypeptide chain.

Henney and Storck (1964) found that polysomes were present in growing and germinating cells of *Neurospora crassa,* a fungus, but were not present in resting and dormant cells which were not synthesizing proteins, suggesting that polysomes are necessary for protein synthesis. Thus it appears that polysomes are included in growing cells and active cells which are involved in protein synthesis but are not present in nonactive cells because there are no mRNA's to hold the individual ribosomes together in a polysome unit. Appel, Davis, and Scott (1967) reported an increase in the number of polysomes in the brains of rats stimulated by light and a decrease in rats deprived of light. Such changes did not occur in the liver, however.

If the synthesis of mRNA is prevented by actinomycin D, polysomes should disappear and be replaced by single ribosomes. This result has been indicated by Staehelin, Wetlstern, and Noll (1963) in rat liver.

If the DNA sites code for protein synthesis via the base sequences in RNA, an important question is "What is the size of the coding unit?" If the codon is by a single base, then there are 4^1 possible results corresponding to A, G, C, and U. If a two letter code is involved, there are 4^2 or 16 possibilities, AA, AG, AC, AU, GA, GG, etc. If a three letter code is offered, 4^3 or 64 different triplets result, AAA, AAG, AAC, etc. If a four letter code is required, 4^4 or 256 quadruplets occur; etc. Because there are 20 common amino acids in nature, a coding procedure of a minimum of three nucleotides for each amino acid has been considered to be most plausible (Crick, 1962; Crick et al., 1961; Crick, 1963; Rich, 1962). Crick (1962) suggested that the message was read in nonoverlapping groups from a fixed point in groups of a fixed size that was probably three, although he did not rule out multiples of three, that there was very little nonsense in the code, and that in general more than one triplet could stand for each amino acid.

Matthaei et al. (1962) developed codes for most of the 20 amino acids by using synthetic polyribonucleotides to synthesize protein. For example, UUU respresents phenylalanine. The code is degenerate, i.e., more than one triplet can code for a single amino acid. For example, two U's and either a C or a U are both considered to be the code for phenylalanine. Table 3 gives the trinucleotide codes based on a summary provided by Harbers, Domagk, and Muller (1968). Sixty-one of the triplets code for

TABLE 3   The 64 Possible Triplet Combinations in Messenger RNA (codons) and Their Corresponding Amino Acids (modified from Harbers, Domagk, and Muller, 1968)

AAA Lysine	CAA Glutamine	GAA Glutamic acid	UAA None
AAG Lysine	CAG Glutamine	GAG Glutamic acid	UAG None
AAC Asparagine	CAC Histidine	GAC Aspartic acid	UAC Tyrosine
AAU Asparagine	CAU Histidine	GAU Aspartic acid	UAU Tyrosine
ACA Threonine	CCA Proline	GCA Alanine	UCA Serine
ACG Threonine	CCG Proline	GCG Alanine	UCG Serine
ACC Threonine	CCC Proline	GCC Alanine	UCC Serine
ACU Threonine	CCU Proline	GCU Alanine	UCU Serine
AGA Arginine	CGA Arginine	GGA Glycine	UGA None
AGG Arginine	CGG Arginine	GGG Glycine	UGG Tryptophan
AGC Serine	CGC Arginine	GGC Glycine	UGC Cysteine
AGU Serine	CGU Arginine	GGU Glycine	UGU Cysteine
AUA Isoleucine	CUA Leucine	GUA Valine	UUA Leucine
AUG Methionine	CUG Leucine	GUG Valine	UUG Leucine
AUC Isoleucine	CUC Leucine	GUC Valine	UUC Phenylalanine
AUU Isoleucine	CUU Leucine	GUU Valine	UUU Phenylalanine

amino acids; the remaining three (UAA, UAG, and UGA) appear to code for the termination of a peptide chain (Caskey et al., 1968). Figure 7 shows the mRNA codon for phenylalanine (UUU) united with the tRNA anticodon (AAA) and the codon (CUA) and anticodon (GAU) for leucine. The anticodon is presumed to be located in the central arm of the cloverleaf and the amino acid, on the CCA arm.

Guthrie and Nomura (1968) and Kondo et al. (1968) reported results which suggested that prior to the beginning of protein synthesis in *E. coli*, 70 S particles separated into the 30 S and 50 S units and there was the formation of an "initiation complex" which involved the binding of a specific tRNA and mRNA to the 30 S ribosomal unit. The specific tRNA binding to this complex carried the amino acid, methionine. The trinucleotide codon in mRNA for methionine is AUG.

Preliminary evidence has suggested that the code is somewhat universal with all species using the same code. For example, synthetic RNA polymers code the same way in a bacterial system as they do in mammalian *in vitro* systems. However, some differences probably do occur. Because of the degeneracy of the code, it is probable that one species may use one of the alternate codes while another species uses a different one.

If we assume that the average size protein in an animal cell is 500 amino acids in length (Watson, 1965), then the DNA code for this molecule would consist of 1500 nucleotide pairs. Each nucleotide pair has a molecular weight (MW) of 660; thus DNA units of MW = $660 \times 1500 = 10^6$ would be required for each protein. The number of different types of proteins can be estimated by dividing the MW of DNA in a cell by 10^6.

The coding results have been obtained by studying the process of protein synthesis in cell-free extracts of *E. coli* from the human colon (Nirenberg, 1963). The bacteria are grown rapidly in suitable nutrients and are harvested by sedimenting them by centrifugation. The cells are gently broken open by grinding them with a fine powder to release the cell sap which contains DNA, mRNA, ribosomes, enzymes, and other components. These extracts are called *cell-free systems* (*in vitro*) and will incorporate amino acids into proteins when they are supplied with energy rich substances (mainly ATP—adenosine triphosphate). The incorporation process is followed by using amino acids containing radioactive carbon (C^{14}).

Even though trinucleotide coding is most accepted, other coding procedures have been suggested. Roberts (1962) indicated that a doublet code would eliminate degeneracy aspects. In this code, a G and a C represent alanine; a G and a U, valine; etc. However, this coding procedure results in ambiguities, e.g., AA codes for lysine and methionine.

Several individuals favored triplet codes which utilize certain doublet aspects. Eck (1963) discussed a symmetrical pattern for the 64 possible combinations of the four nucleotides taken three at a time. He suggested 32 pairs in which the second member of each pair was identical to the first except that in one position a purine is replaced by the other pyrimidine. This symmetrical pattern was consistent with the reported triplet codes, allowed for prediction of amino acids corresponding to unidentified triplets, and suggested a structural basis for tRNA specificity. These triplets might be called a "two-and-a-half-letter" code in that one nucleotide could contain either of the purines (or either of the pyrimidines) and still be "recognized" by one tRNA. Thus there would be 64 possible triplets on mRNA but only 32 different tRNA's.

Jukes (1963) suggested a "modified doublet" coding procedure which might be appropriately entitled a "greater than two but less than three letter" code. It was postulated that in each triplet there is a pivotal base which is subject to change without altering the coding function of the triplet. The pivotal base could not be in the third position nor could it be a G; it was U in 11 of the codes containing this base.

Certain observations have posed a problem for these genetic coding attempts. If synthetic RNA containing only uracil bases (poly U) is used as a template and is combined with ribosomes from *E. coli*, plus other necessary constituents, one would expect that phenylalanine would be produced. However, phenylalanine incorporation is markedly dependent on magnesium ion concentration; when the concentration was increased, incorporation increased for amino acids (e.g., isoleucine) which were not normally "coded" for by poly U, and incorporation of phenylalanine was reduced (Hechter and Halkerston, 1965). These results show the importance of internal environmental agents in protein synthesis.

DNA Regulation

In the behavior of organisms there is the need for some system to inhibit, control, or regulate others so that organized behavior will result. If no regulation were involved, behavior would be chaotic. At the gene level, it has been suggested by Jacob and Monod (1961), and others, that some DNA sites control the RNA synthesis of other DNA sites and, thereby, the overall functional activity of the cells.

Busch et al. (1964) suggested five genetic operating groups or poly-operons involved in cell functions. These are:

1. Basic cell metabolism
2. A specialized cell function
3. Cell growth
4. Cell division
5. Abnormal cell growth, e.g., cancer

All, or most, cells of an organism could share in common the components of the first group which provides the basic elements necessary for the fundamental metabolism of the cell; however, the function for the other four would vary from cell to cell. In recent years some behavioral scientists have suggested that another genetic operating group should be present in brain cells, viz., DNA sites which would provide unique contributions in learning behavior.* Although this is a popular conception today, the results relative to this point are inconclusive.

For regulation of these cell functions one would expect that some DNA's would be active in RNA synthesis whereas others would be inert. This appears to be the case. All cell nuclei contain the same genetic information but not all DNA sites can be active in synthesizing RNA. If all were active, cell differentiation would not occur and every cell would be the same. Thus two different types of DNA (in terms of functional activity) are to be expected.

A basic question in biology today concerns the mechanism whereby genes are turned on and off. DNA activity is assumed to be highly cell specific (Markert, 1965); some genes function only in certain cells, and then only when the cell has reached a suitable stage in development. Presumably brain cells differ from other cells in that in the former, certain DNA sites are capable of synthesizing RNA whereas these sites are repressed permanently in the latter. The liver must have DNA sites which are turned off in brain DNA but which are functional in the liver. One would expect that similar results would occur for the DNA of other tissues also. A good example of this specificity is the gene for tyrosinase which is active only in one cell, a melanocyte, at terminal stages of differentiation. This gene is present in other cells but is silent. A number of other examples can be cited: karatin is synthesized in epithelial cells, thyroglobulin in thyroid cells, actin and myosin in muscle cells, insulin in pancreatic islets, etc.

If DNA regulation occurs, one would expect that RNA from different tissues would not be the same. Dingman and Sporn (1962) indicated that the cytoplasmic RNA of the microsomes of rat brain did not differ from that in rat liver, but differences were present in nuclear RNA of the two tissues. Furthermore, the nuclear RNA of young rats was different from that of adult rats. They were not able to indicate whether the differences were in primary, secondary, or tertiary structure. In another study, they found that DNA-histone complexing was different in various cells. The proportion of histone complexed with DNA was greater in erythrocytes than it was in brain and liver (Sporn and Dingman, 1963).

Miyagi, Kohl, and Flickinger (1967) used DNA-RNA hybridization procedures (see Chapter 7) in an attempt to distinguish RNA species in the liver and kidney of adult chickens. Their results suggested the presence

---

* This function might be considered under (2), a specialized cell function for brain cells.

of RNA species in the liver which were qualitatively different than those in the kidney and the presence of adult-like RNA in the liver of two-and-a-half-day-old chick embryos.

An interesting regulatory event has been shown with a variety of mammalian species including the human female (Grumbach, Morishima, and Taylor, 1963). One of the X sex chromosomes in some cells undergoes a change of state which first becomes manifest during early development. This X chromosome becomes highly condensed and is characterized by relative genetic inactivity. The other X chromosome in the female and the single X in the male are genetically active and in noncondensed form. Grumbach, Morishima, and Taylor suggested an induction model patterned on the Jacob-Monod approach to account for these events.

In 1961 Jacob and Monod introduced a gene regulation model which has had a tremendous impact on thinking in Molecular Biology. They postulated three types of genes: regulator, operator, and structural. The operator and structural genes are linked together as an operational unit (an operon). These two types of genes in an operon are adjacent, with the operator gene functioning to turn the operon on or off. Thus the operator gene determines whether the code in DNA will be transcribed into RNA or not. There is one operator gene for one or more structural genes. The operator gene is controlled by another gene at some distance from the operon unit, the regulator gene. The operon is in an on state. A regulator DNA codes for a repressor substance (probably a protein) which under certain conditions can complex with the operator gene; this event inactivates the structural gene and transcription of mRNA is prevented. If an inducer is present it will attach to the repressor, thereby removing the repressor and allowing RNA to be synthesized. For example, if β-galactoside is present, it attaches to the repressor molecule on the operator gene and removes the repressor substances. This removal switches on the synthesis of RNA for β-galactosidose by the structural gene. In some cases, however, the inducer substance combines with products synthesized as a result of the function of structural genes. For example, with excessive amounts of tryptophan available in *E. coli,* the inducer substance binds to this amino acid; thus this event prevents the inducer from removing the repressor from the operator gene; thus the operon remains in the off position, and there is no RNA synthesized which codes for tryptophan synthetase, an enzyme required for the synthesis of tryptophan. In this model there is a regulator-operator-structural gene unit for each biosynthetic pathway. Figure 9 shows a diagram of this model.

Portions of the Jacob-Monod model have been confirmed with bacteria and bacterial viruses. For example, Ptashne (1967) and Bretscher (1968) reported the isolation of repressor substances (proteins) which bound specifically to DNA sites, thereby blocking DNA transcription into mRNA.

Tomkins et al. (1969) suggested that the Jacob-Monod model was

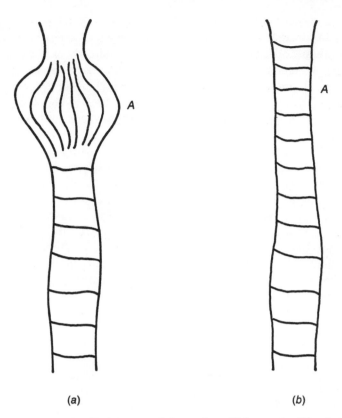

(a)   (b)

FIGURE 9   A simplified diagram of the regulator (Re)–operator (O)– structural (S) DNA scheme. *a.* Components in system; *b.* enzyme induction-inducer (I) removes repressor (Rp) by binding to it and allows the turning on of structural DNA's; *c.* enzyme repression-corepressor (C) binds to inducer and attachment of inducer to repressor is prevented.

suitable for gene regulation in bacteria but not in animal cells. They pointed out that numerous studies suggest that large portions of the genome of animals cells are transcribed continuously into RNA in the nucleus, but only a small fraction of these RNA's reach the cytoplasm. Thus a great variety of RNA molecules must be synthesized and destroyed, never functioning as mRNA's although they have the potential to do so. For animal cells, they stated that the repressor substance acts not at the DNA level to prevent RNA synthesis but at the RNA level itself. When the repressor combines with mRNA, degradation of the RNA results. If an inducer is present, the repressor is inactivated and can not combine with mRNA, thereby allowing protein synthesis to occur.

Another recent paper (Britten and Davidson, 1969) suggested a modification of the Jacob-Monod model. Britten and Davidson hypothesized a fourth type of gene called a *sensor*. This gene is the binding site for agents which lead ultimately to RNA and protein synthesis. For example, hormones induce synthesis by binding to sensor DNA sites (directly or via intermediary molecules). When inducers bind to sensor genes, an adjacent integrator gene is turned on to produce activator RNA. The sensor-integrator genes are the equivalent of the regulator gene of Jacob and Monod. The activator RNA forms a sequence specific complex with a receptor gene that is adjacent to a producer gene. The receptor and producer genes are the equivalent of the Jacob-Monod operator and structural genes, respectively. The activator RNA in complexing with the receptor gene switches on RNA synthesis by the producer DNA.

The major differences in this last model and that of Jacob and Monod are the number of types of genes (three for Jacob and Monod; four for Britten and Davidson) and the site of influence of inducers (on the repressor molecule for Jacob and Monod; at sensor DNA for Britten and Davidson). Both models, however, regulate amounts of RNA at the gene level in contrast to that by Tomkins et al. (1969) which regulates the amount of RNA after synthesis.

There are possibly a number of types of regulation at the gene level. Different types of gene regulation may occur in different cells or in the same cell at different times. Likewise, one should keep in mind that these models refer to intracellular events; obviously extracellular regulation occurs also. Such regulation may take a number of forms. Probably one of the more important ones in mammals is the contribution of the reticular formation of the brain stem in affecting the tonicity of other neural structures and possibly the upper portion in the thalamus serving as an integrating center (Penfield, 1960).

An understanding of the overall regulatory functioning is a basic problem for biology today but little is actually known concerning this function. Possibly with the tremendous advances at the molecular level, new insights may be suggested at higher levels and rapid development may ensue on this important problem.

Chapter 5

RESEARCH CONCERNED WITH
DNA REGULATION

There are two aspects of the DNA Complex which will provide a structure for this chapter and the following four: the turning on and off of genes (DNA regulation) and the products of DNA activity. In this chapter the research results relative to DNA regulation will be discussed. These research efforts have been conducted mainly within the domain of Molecular Biology.

In Chapter 4 it was indicated that gene regulation was necessary to allow cellular differentiation and organized behavior to occur. For convenience, studies which are related to this aspect will be discussed under five headings which are not mutually exclusive: active and inactive DNA sites, hormonal control, enzyme induction and repression, histone effects, and nuclear proteins related to behavior.

Active and Inactive DNA Sites

All DNA sites are not active in synthesizing RNA, even though all cell nuclei contain the same basic genetic code. For regulation of DNA sites one would expect that some DNA's would be active whereas others would be inert. This appears to be the case. Thus, in terms of functional activity, two different types of DNA are to be expected: active and inactive. Of the functionally active DNA, one could be active for the synthesis of DNA whereas other DNA would show its activity in RNA synthesis. It is this latter type of DNA which is of most concern in this book. In the brains of higher animals very little DNA synthesis occurs, but RNA synthesis is quite prominent during behavior.

Swift (1962) concluded from his studies on DNA in certain species of flies that there are two types of DNA: one which is constant in amount from cell to cell and another varying in amount with particular cell type and the particular stage of ontogeny. Likewise, Bendich, Russell, and Brown (1953) found two functional types of DNA in growing rat tissue, one showing a higher turnover* than the other. Sampson et al. (1963) reported two DNA fractions in plants: a high molecular weight stable fraction and a low molecular weight fraction showing a relatively rapid rate of turnover. The proportion of the two forms varied with physiological state and with the type of tissue. Male germinal tissue contained negligible amounts of the low molecular weight form. Growing regions of root and leaf had as much as 20 percent of their total DNA in the low molecular weight form. Dormant embryos in wheat seeds had about 10 percent, but the amount increased sharply upon induction of germination.

Sampson et al. indicated that in respect to the properties investigated the high molecular weight stable form behaved typically for genetic material.

* As measured by incorporation of radioactively labelled precursors into DNA, an event which indicates DNA synthesis.

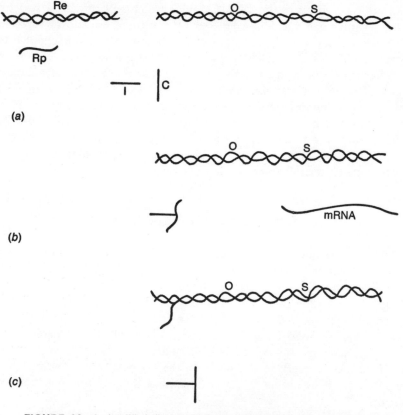

FIGURE 10 A simplified diagram indicating chromosomal puffing at *A* in the giant salivary glands of a fly, *Drosophila*. *a.* Puffed chromosome; *b.* non-puffed chromosome.

However, the low molecular weight active DNA appeared to be performing a physiological role via RNA synthesis. Likewise, Frenster, Allfrey, and Mirsky (1963) isolated active and repressed DNA fractions from calf thymus chromatin.

In some insects, e.g., *Drosophila,* a fruit fly, the salivary glands contain very large cells with large nuclei in which giant chromosomes can be observed with a microscope. Investigations have shown puffing at specific sites in these chromosomes. These sites are assumed to indicate active DNA; the sites of activity appear to be at different loci in different tissues and at different loci in the same tissue at different times (Beermann and Clever, 1964; Harbers, Domagk, and Muller, 1968).

That these puffs represent DNA becoming active in the synthesis of RNA has been indicated by studies using actinomycin-D, an inhibitor of RNA synthesis. When this antibiotic is administered prior to the time for puffing, the puffs do not appear. A diagram of puffing is shown in Figure 10.

Hormonal Control

A number of studies have found an increase in RNA synthesis following the administration of hormones. Kenney and Kull (1963) found that hydrocortisone increased the rate of synthesis of liver nuclear RNA during the induction of an enzyme, tyrosine transaminase, in adrenalectomized rats. Wicks and Kenny (1964) reported that testosterone injections in castrated rats resulted in a two to threefold increase in RNA synthesis in the seminal vesicle within 50 minutes after injection. Insulin led to an increase in RNA in rat diaphragm (Wool and Munro, 1963). Leslie (1955) and Brown and Roll (1955) reported increases in RNA and RNA/DNA ratios following the administration of various hormones. Olson (1964) found that actinomycin D inhibited vitamin K induced formation of prothrombin in chicks deficient in vitamin K; inhibition of RNA synthesis also occurred. He suggested that vitamin K and all fat-soluble vitamins operate to control the synthesis of specific proteins and enzymes. Other individuals (Karlson, 1962; Schneiderman and Gilbert, 1964; Harbers, Domagk, and Muller, 1968) reported that hormones such as ecdysone cause chromosomal puffing in insects, resulting in rapid RNA synthesis in puffed regions (Figure 10). A small dose of this hormone leads within 30 minutes to puffing at a specific site on one chromosome, and 30 minutes later to the formation of another puff in a second chromosome. These events are followed by a chain of secondary puffing reactions. Clever (1964) reported a systematic study concerning the effects of ecdysone alone, or with specific antibiotics, on the degree of puffing at specific sites in the giant chromosome of an insect, *Diptera.* Table 4 summarizes the results for one chromosomal site. Under control conditions, most animals show no puffing (0); a few show little puffing (1). Using ecdysone

TABLE 4   Effects of Various Conditions on Degree of Puffing at a Specific Site in a Giant Chromosome of *Diptera* (after Clever, 1964)

Treatment	No. of Animals in class			
	0	*1*	*2*	*3*
Control (Untreated)	15	5	0	0
Ecdysone	0	0	5	14
Puromycin + Ecdysone	0	0	4	9
Actinomycin + Ecdysone	11	3	1	0
Ecdysone + Actinomycin	8	3	0	0

The classes refer to degree of puffing from 0 (no puff) to 3 (maximal puff).

most animals have maximal puffing (3), with a few having moderate puffing (2). When puromycin (a protein synthesis inhibitor) is combined with ecdysone, the results are as with ecdysone alone. When ecdysone is used along with actinomycin, the results are the same as in the control condition, irrespective of the order in which the two are administered. Presumably the antibiotic prevents RNA synthesis and precludes the formation of puffs.

Bonner et al. (1968) stated that numerous hormones effect rapid RNA synthesis. However, the addition of the hormones directly to isolated chromatin did not increase RNA synthesis. Therefore, the effect of the hormone must be mediated by some substance or substances present in the nucleus but not in the purified chromatin.

The period following the administration of a hormone and the appearance of an effect on various portions of the DNA complex varies from about two minutes to 24 hours. Talwar, Sharma, and Gupta (1968) suggested that this variation resulted because each hormone has an early primary effect but also other effects of a secondary nature which develop later.

Harbers, Domagk, and Muller (1968) summarized research on hormonal stimulation of RNA synthesis in specific tissues (Table 5). They concluded that there was an increase first in the formation of rRNA and later in specific mRNA's.

Hamilton (1968) provided a comprehensive review of the literature concerning the physiological effects of estrogen in the mammalian uterus. He concluded that although the precise molecular mechanisms are almost totally unknown, there is suggested the following sequence of intracellular events:

1. binding of the hormone to the chromatin in the nucleus,
2. stimulation of chromosomal and rRNA synthesis in conjunction with chromosomal and nucleolar activity,
3. acceleration of the rate of formation of ribosomal precursor particles,

TABLE 5 Effective Stimulation (+) of Nuclear RNA Synthesis *in vivo* or in Isolated Nuclei by Various Hormones (after Harbers, Domagk, and Muller, 1968)

Hormone	Species	Target Tissue	Stimulatory Effect on the Incorporation into RNA
ACTH	Rat	Adrenals	+
Cortisol	Rat	Liver	+ Stimulation of RNA coding for tyrosine transaminase
Ecdysone	Blowfly (Calliphora)	Epidermis and whole larvae	+ Stimulation of RNA coding for DOPA decarboxylase
Estrogen	Rat	Uterus	+ mRNA from uterus of estrogen-treated rats, applied intrauterine to ovarectomized animals caused morphological changes similar to those induced directly by estrogens
Insulin	Rat	Diaphragm	+
Testosterone	Rat	Prostate Seminal vesicles	+
Thyroid hormones	Rat Tadpole Tadpole	Liver Liver Tail	+
TSH	Sheep	Thyroid	+
Plant hormones and regulators (auxin, cytokinins, gibberellins)	Various plants	—	+

4. acceleration of transport of ribosomal precursor particles with attached mRNA to the cytoplasm, and

5. accumulation in the cytoplasm of new polysomes having different amino acid incorporating properties than the old ones.

Hamilton maintained that the effects of estrogen on cytoplasmic protein was indirect, with the effect on RNA synthesis resulting in a regulation of the rate and amount of protein synthesis.

O'Malley et al. (1968) provided results which suggested that estrogen may stimulate the synthesis of RNA in the oviduct of the immature chick. A fivefold increase or more in 4 S RNA occurred in both nuclear and cytoplasmic fractions.

Enzyme Induction and Repression

Enzyme induction and repression events were instrumental in the formulation of the Jacob-Monod regulator-operator-structural DNA model (1961). The addition of an inducer such as thio-methyl-β-galactoside to a strain of *E. coli* stimulates the synthesis of the enzyme β-D galactosidase. The genetic region responsible for β-galactosidase in *E. Coli* was determined in genetic experiments. This region is referred to as the *lac* region. Hayashi et al. (1963) added thio-methyl-β-D-galactoside to an *E. coli* medium and obtained increased amounts of β-galactosidase; they also found increased amounts of RNA which were complementary to the DNA of the *lac* region. These results supported the Jacob-Monod model of inducer action at the gene level. Other work (Pollard, 1964) indicated that the time of onset of enzyme activity after induction was about three minutes.

Watson (1965) stated that in each *E. coli* cell growing in the presence of β-galactosides, there are 30 to 50 RNA's specific for β-galactosidase and approximately 3000 β-galactosidase molecules. The ribosomes constitute about 25 percent of the cell mass. In contrast, when β-galactosides are metabolized and disappear, no enzyme will be synthesized, and the average cell will contain fewer than one mRNA specific for β-galactosidase synthesis; and the ribosome content drops to approximately 5 to 10 percent of the cell mass.

The presence of a specific chemical may turn off the synthesis of a specific enzyme (enzyme repression). For example, if there is no tryptophan (an amino acid) in the culture media of *E. coli,* these bacteria synthesize the different enzymes which control the various steps in the synthesis of tryptophan. If tryptophan is present, however, the enzymes are not produced.

Filner, Wray, and Varner (1969) discussed induced enzyme synthesis in plants. Many cases which appear to be enzyme induction are available, e.g., in barley cells and in peanuts. Large fluctuations of enzyme activities occur in plants in response to such factors as hormones, light, dark, air, water, and development (time).

Enzyme induction events have also been incorporated in a number of learning models dealing with higher animals. These models will be discussed in Chapter 11.

Histone Effects

Histones complexed with DNA have been shown to affect DNA activity. Huang and Bonner (1962) found with pea embryo chromatin that when the protein fraction, histone, was removed, the rate of RNA synthesis increased fivefold. Further work (Bonner and Huang, 1962a) showed that the chromatin contained 80 percent DNA bound to histone and 20 percent of DNA free of histones. They suggested that the function of histone was to bind DNA and block the transfer of "information" from DNA. Bonner and Huang (1962b) also discussed the removal of inhibition following a decrease of histones in certain plants in transition from the vegetative to the flowering state. In these plants induction of flowering is brought about by exposure of a single leaf of the plant to a single long night. The leaf then sends a signal to the bud, the bud becomes transformed, and some 48 hours after the beginning of the long night, histological visible symptoms of floral differentiation begin to make themselves apparent. It was shown that shortly after receiving the flowering message by the bud, and before any visible signs of differentiation were apparent, a sharp drop in histone content of the bud occurred. This was followed by a dramatic increase in RNA concentration in the same cells.

Bonner and Huang (1964) reported further experiments indicating inhibition of gene functions by histones. They found that the very lysine rich fraction was the most effective for inhibiting RNA synthesis. This fraction contained more of the amino acid proline than the other fractions. The physical characteristics of the fractions appeared to differ also: the temperature required to denature DNA into single strands was greatest for the most inhibitive, and all nucleohistone fractions required higher temperatures than did DNA alone.

Marushige and Bonner (1966) showed that rat liver chromatin is 20 percent as active in RNA synthesis as deproteinized DNA prepared from it. Furthermore, RNA synthesized using chromatin differs in base composition from that synthesized using histone-free DNA as template. Selective removal of histone protein, but not of nonhistone protein, increased the amount of RNA synthesis to that of deproteinized DNA. By progressive removal of histones from pea bud chromatin, Bonner, Huang, and Gilden (1963) and Bonner and Huang (1966) were able to demonstrate that synthesis of RNA involved in the synthesis of pea seed globulin was accompanied by the removal of a particular histone fraction.

It was reported (Hnilica and Bess, 1965; Bonner, 1965; Bonner and Huang, 1966) that a previously unrecognized type of RNA, chromosomal RNA, is bound to a nonhistone protein associated with DNA in chromatin. This RNA plays an important role in DNA regulation according to the Bonner and Huang (1966) summary of the role of histones: "...histones are repressors of the transcription of DNA by RNA polymerase...histone molecules in pea bud nucleohistone are grouped together in larger entities, of which the molecules we know by acid extraction are subunits..." (p. 31). Some of the larger units are associated with a nonhistone protein-RNA complex through hydrogen bonds. It was suggested that the function of RNA in these structures is "to detect the gene which is to be repressed, detection being based upon a base-pairing strategy...the RNA in question seeks its complementary operator, the associated histone then complexing with DNA of the adjacent structural gene" (p. 32). It appears that chromosomal RNA performs a function opposite to that of the activator RNA of Britten and Davidson (1969). Chromosomal RNA complexes with operator DNA so as to allow the associated histones to attach to structural DNA and inhibit the RNA synthesis. Activator RNA complexes with receptor DNA (equivalent to operator DNA) to switch on the RNA synthesis by producer DNA (same as structural DNA).

In a recent article concerning the role of histones, Bonner et al. (1968) provided an excellent summary on this matter. Template activity for RNA synthesis catalyzed by RNA polymerase is less for DNA in isolated chromatin than for deproteined DNA. This was true for various plant and animal chromatins. The percent template activity (relative to deproteined DNA) varied from 6 percent in pea vegetative bud to 32 percent in pea growing cotyledon. Some cells from rats, humans, cows, and sea urchins fell within these extremes. Thus deproteinized DNA showed 3 to 16 times as much template activity as did the chromatins. (See Table 2, Chapter 3.) DNA-RNA hybridization studies (see Chapter 7) indicated that a restricted portion of chromatin DNA functions in RNA synthesis. For example, RNA synthesized *in vitro* from DNA hybridizes with approximately 15 times more denatured DNA than does RNA synthesized from thymus chromatin *in vitro*. Bonner et al. concluded that histones were the agents responsible for the restriction of chromosomal activity in RNA synthesis.

They reported that the specificity for the interaction between DNA and histones in regulating gene activity was via a chromosomal RNA. This RNA contains 40 to 60 nucleotides, has a sedimentation coefficient of 3.2 S, includes a relatively high content (5 to 25 percent) of dihydrouridylic acid, and hybridizes with slightly over 5 percent of DNA. They stated that chromosomal RNA was exceedingly heterogeneous in base sequences and consisted of many species of RNA, each represented but a small number of times.

Other individuals have stressed the importance of histones in regulating RNA synthesis. Izawa, Allfrey, and Mirsky (1963), working with giant chromosomes of amphibian oocytes (eggs), found that the addition of thymus gland histones to isolated nuclei inhibited RNA synthesis. These histones consisted of two parts: one rich in arginine and the other in lysine. The arginine portion was an effective inhibitor of RNA synthesis whereas the lysine portion produced a weak inhibition or was ineffectual. [This result was in contrast to those of Bonner and Huang (1964) discussed above, but the systems of the two groups were completely different.] They also found that active chromosomal material in the oocytes contained less histone complexed with DNA than did the repressed material of calf thymus lymphocytes (Frenster, Allfrey, and Mirsky, 1963); however, Swift (1964) reported via histochemical studies that in puffed regions of giant chromosomes where RNA synthesis is intense, the histone content is about the same as in the remainder of the chromosome. Izawa, Allfrey, and Mirsky revealed further that the DNA active in RNA synthesis was mainly in diffuse, extended form rather than in condensed, compact masses (Littau et al., 1964).

Allfrey et al. (1966) summarized various reports which show a relationship between acetylation of histone and RNA synthesis in chromatin. Chemical acetylation *in vitro* of arginine-rich histones reduced their capacity to inhibit RNA synthesis but did not prevent the modified histones from combining with DNA. Uptake of C^{14}-acetate *in vivo* by histones was pronounced in the arginine-rich fraction and proceeded independently of protein synthesis. Incorporation of H^3-uridine into RNA was most rapid in regions of chromosomes which showed puffing; this result was paralleled by the distribution of incorporated C^{14}-acetate. A positive relationship was demonstrated between the rate of RNA synthesis and acetate incorporation in isolated nuclei of various tissues. Corticosteroids can produce changes in the rate of RNA synthesis in certain tissues, a reduction being produced in thymus but an increase in liver; parallel changes are produced by incorporation of C^{14}-acetate in these tissues. Allfrey et al. (1966) concluded that a change in the structure of chromatin—brought about by, or coincident with, acetylation of histones—was a necessary prerequisite to the synthesis of new RNA's at previously repressed gene loci.

Pogo, Allfrey, and Mirsky (1966) investigated the effects of phytohemagglutinin (PHA—a protein fraction derived from the red kidney bean) on human lymphocytes. These cells when maintained in tissue culture rarely divide. However, in the presence of PHA, the cells increase their metabolic activity, enlarge, and divide. During the process, increased histone acetylation and synthesis of RNA and protein occurred; the increase of acetylation appeared to precede the increase in RNA synthesis. These results suggested that inactive gene loci were called into play. The increased activity of the DNA could be detected long before the cells enlarged and went into mitosis. Their results were consistent with the view that histone acetylation signals a change in the fine structure of the chromatin which leads to increased RNA synthesis.

Allfrey et al. (1966) also reported that histone acetylation occurs in the giant salivary gland chromosomes of *Chironomus thummi* when RNA synthesis is active. They reached similar conclusions regarding phosphorylation of histones, which takes place after synthesis of these proteins. Phosphoprotein appeared to be more concentrated in active chromatin than in condensed chromatin and it interacted with histones *in vitro* to diminish the inhibitory effects of histones on RNA synthesis. Phosphorylation of nuclear proteins appeared not to be dependent on RNA synthesis, since agents which block RNA synthesis did not inhibit phosphorylation. Allfrey et al. suggested that phosphorylation of chromosome associated proteins influenced DNA-histone interactions and led to a shift from the condensed to the diffuse state, while dephosphorylation would lead to tight coiling of the DNA-histone complex.

In a recent report on histone phosphorylation, Gutierez and Hnilica (1967) concluded that while all histone fractions are capable of phosphorylation, there is much variation between fractions. Histone phosphorylation appeared to be tissue specific and more rapid in regenerating or neoplastic tissues than in well-differentiated tissues. Active phosphorylation of histones appeared to be associated with metabolic functions of the cells and not with growth, since the relative extent of P^{32} incorporation decreased with increasing mitotic rates.

Work by Allfrey and Mirsky (1964) showed agreement with the Bonner and Huang (1964) findings that the very lysine-rich fraction was most inhibitive when the systems were similar. They also reported that inactive DNA was readily removed from isolated nuclei by deoxyribonuclease (DNase) but that active DNA required greater amounts of DNase or more prolonged incubation to be eliminated.

Busch (1965) indicated that histones cause a cessation of cell division and embryonic development in frogs' eggs. Billen and Hnilica (1964) reported that DNA synthesis was suppressed completely when a high concentration of calf thymus histones were added to an *in vitro* system of *E. coli* DNA polymerase and calf thymus DNA. More than 50 percent inhibition was obtained when the weight of DNA to histone was 2:1. The inhibition was accompanied by formation of a fibrous nucleohistone precipitate and loss of ultraviolet absorbing material, which paralleled the decline in DNA synthesis. Increasing the concentration of DNA brought about an almost complete reversal of the inhibition. Presumably the loss of DNA template activity was caused by the precipitation of DNA. Any agent precipitating DNA, and thus altering its overall structure, would result in loss of DNA or RNA synthesis.

Goodwin and Sizer (1965) reported results which suggested that specificity of histones for a regulatory role resided in concentration rather than in type of histone present. They found that enzymatic activity in cultures of embryonic chick brain tissue was stimulated by low concentrations of histones but was repressed at higher concentrations. This control was shown to operate by a modification of protein synthesis, presumably by changes in mRNA synthesis.

Neidle and Waelsch (1964) investigated the specificity of histones relative to species and tissues. They reported that within a given species (rat, mouse, guinea pig, rabbit), the histones of the brain, liver, and kidney were grossly indistinguishable. Differences were observed between species. In immature rats the histones of brain and liver were similar but differed from those in the adult rat.

The above results suggest that histones have an important regulatory role in RNA synthesis. Either quantitative or qualitative changes (acetylation, phosphorylation) in histones appear to affect RNA synthesis. The exact mechanisms, however, are still obscure.

Although the emphasis of this section has been on the possible role of histone in RNA synthesis, there are acidic proteins in the nucleus which need to be discussed. Their importance metabolically is evident from their high rates of turnover (Busch et al., 1963). The acidic proteins and histones are approximately equal in amount, each comprising approximately 20–25 percent of the total dry weight of the nucleus in the rat liver. Busch et al. (1963) hypothesized that DNA and the acidic proteins compete for linkage with histones and that loss of histones from linkage with DNA to the acidic proteins would free DNA for synthesis functions. Thus the acidic proteins may be involved in the regulation of gene activity (Markert and Ursprung, 1963; Busch et al., 1963).

An acidic protein which is found only in the brain of some higher organisms has been reported recently. This S-100 protein has been considered by some individuals, e.g., Hydén, to be of importance in learning events.

Nuclear Proteins Related to Behavior

Based on the results discussed in the previous sections of this chapter, it appears that prior to the occurrence of RNA synthesis, specific DNA sites puff to effect this synthesis. Various agents such as hormones precipitate this result. Possibly during behavioral events, e.g., learning, puffing is initiated also. The previous sections indicate that some chemicals such as histone function to reduce the template activity of DNA. Thus during learning, quantitative or qualitative changes in histone may occur so that DNA puffing and RNA synthesis may result.

To investigate the possibility that histone changes may regulate RNA synthesis during behavioral events, a research program was initiated in the Molecular Psychobiology Laboratory of York University to attempt to relate qualitative and quantitative changes in histones (and other nuclear proteins) to learning events.

At this time only preliminary experiments with shock avoidance conditioning have been conducted. Extraction and analytic procedure suitable with calf thymus tissue appeared to be adequate with rat brain tissue as well. However, the results in these preliminary experiments indicated that the yield and specific activity* of histones were too low to allow for comparisons between learning and nonlearning rats. At the present time the procedures are being improved. Ultimately, this research project relating histones and other nuclear proteins to learning phenomena should provide some basic results.

* Amount of labelled precursors incorporated within a specific amount of histones.

Chapter 6

RESEARCH CONCERNED WITH THE PRODUCTS OF DNA ACTIVITY

The research discussed in this chapter, and the following three, has occurred within Molecular Psychobiology. In this newly developed area there are two divergent orientations which are reflected in the various research programs. There is the molecular biological orientation or direct approach which attacks the problem of the relationship between behavior and neurochemistry by varying behavior and analyzing the resulting changes in the brain of the organism. The research discussed in this chapter and the next is of this nature. The other orientation, or indirect approach, is the behavioral or psychological one which involves the administration of specific chemicals and the analysis of behavioral changes which result (Chapters 8 and 9). Both of these approaches are valuable, but the author believes that the first approach is the most efficacious and that the second should be a supplement to the first.

There has been much research which indicates conclusively that RNA and protein changes occur during behavioral events (Gaito, 1966). In general when the behavioral aspect (learning, sensory stimulation, motor activity) is of a mild nature, increases result in RNA and protein. When the behavioral event is drastic or continued for excessive periods, decrements occur.

Although there is no doubt that a relationship exists between RNA-protein and behavior, the exact nature of this relationship is not clear. In attempting to understand the relationship of RNA-protein changes with behavior (and specifically, learning events), it is necessary for the reader to recognize a number of possible interpretations or relationships, i.e., these changes are either primary effects, secondary effects, or parallel effects.

1. PRIMARY EFFECTS

The preliminary phases during the overall situation may lead to RNA and protein synthesis within the brain which, directly or via other molecules, allow behavioral events to occur. This scheme is as follows:

$$RNA \longrightarrow Protein \longrightarrow Behavior$$

In this case the behavioral events of concern are dependent upon the presence of RNA and protein changes, presumably synthesis. If the synthesis of either RNA or protein were prevented, behavioral events could not occur.

Let us concentrate on learning events as the behavior of concern. One can consider RNA and protein changes as having a primary effect on both acquisition (learning) and retention (memory), on acquisition alone, or on retention only. (For further discussion on these possibilities, see Gaito, 1969).

With this type of relationship in mind, one might attempt to improve behavior by administering RNA or chemicals which increase RNA levels. Or one might administer actinomycin-D or other chemicals which reduce RNA amounts and determine if behavioral decrements occur. Likewise, the investigator could use chemicals which affect protein and note the effect on behavior. These procedures have been popular ones and are discussed in Chapters 8 and 9.

2. SECONDARY EFFECTS

It is possible that the behavioral events precede the RNA and protein changes, viz.,

$$Behavior \longrightarrow RNA \longrightarrow Protein$$

If RNA or protein synthesis were prevented, behavior (e.g., learning aspects) would proceed without detriment in this scheme.

3. PARALLEL EFFECTS

A final possibility is

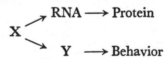

In this case some nervous system functioning (X) is leading both to RNA and protein synthesis as well as to Y, an unspecified component which is necessary before specific behavioral events can occur. The RNA-protein and behavioral events are parallel effects which are not dependent upon one another but follow from a common background (X). Interference with RNA or protein synthesis would have no effect on behavior directly; however, RNA and protein changes might indirectly affect Y, which effect might then be transmitted to behavior.

Researchers who ascribe an important role for RNA and protein within learning events maintain the primary effects position; however, the burden of proof is upon them to exclude the secondary and parallel effects possibilities when relating RNA-protein changes to learning (or to other behaviors). In this and the following chapters the reader might question whether such exclusion has occurred when investigators maintain an important role for RNA and/or protein based on such changes.

Holger Hydén of Sweden has been the pioneer in attempting to relate RNA to behavioral events. He demonstrated that RNA and proteins are produced in the nerve cells at a rate which follows the neuronal activity (1959, 1961). He suggested that the nerve cell fulfills its function under a steady and rapidly changing production of proteins, with the RNA as the governing molecule; he hypothesized that RNA has a unique function in learning events. (See Chapter 11.)

Hydén and Egyhazi (1962) exposed young rats to a situation in which they had to balance on a wire to reach a platform where food was located. The only way for the animals to satisfy their hunger was to accomplish this task. Most rats took an average of four days of 45 minute sessions to balance on the wire the total distance to the food platform; some rats accomplished this skill during the first session. Four days later they were visiting the platform an average of 20 times within the allotted 45 minutes. The animals then were sacrificed and Deiter nerve cells (giant cells) from the vestibular nucleus of the lower brain stem were dissected. Control rats remained in the home cages. Functional controls were provided by rotating animals through 120° horizontally and 30° vertically with 30 turns per minute for two periods of 25 minutes on each of four days. The average amount of RNA per cell was: controls, 683 micromicrograms ($\mu\mu$g); functional controls, 722 $\mu\mu$g; and experimentals, 751 $\mu\mu$g. From these results there is the implication that a greater amount of stimulation was involved in the learning task than for the functional controls. However, the qualitatively different stimulation for the functional controls may be responsible for the difference.

Hydén and Egyhazi also investigated the base composition of the RNA extracted. They found that the cytoplasmic RNA of controls was no different than that of the experimentals (Table 6). In nuclear RNA, there were

TABLE 6 Mean Percent Base Composition of Cytoplasmic RNA from Deiter Cells for Experimental and Control Rats (Hydén and Egyhazi, 1962)

Bases	Controls	Experimentals
Adenine	20.5	20.9
Guanine	33.7	34.0
Cytosine	27.4	26.8
Uracil	18.4	18.3

TABLE 7   Mean Percent Base Composition of Nuclear RNA from Deiter Cells for Experimental and Control Rats (Hydén and Egyhazi, 1962)

Bases	Controls	Functional Controls	Experimentals
Adenine	21.4	21.3	24.1
Guanine	26.2	25.7	26.7
Cytosine	31.9	31.3	31.0
Uracil	20.5	21.7	18.2

TABLE 8   Mean Percent Base Composition of Glial and Nerve Cell RNA of Control Rats (Hydén and Egyhazi, 1963)

Bases	Nerve Cells	Glia
Adenine	20.5	25.3
Guanine	33.7	29.0
Cytosine	27.4	26.5
Uracil	18.4	19.2

significant differences in base amounts. In the experimental group there was a greater amount of A and lesser amounts of U than in the other two groups leading to differences in the A/U ratio (Table 7). The authors maintained that the results indicated the change of RNA bases during learning.

Hydén and Egyhazi (1963) did a similar experiment but were concerned mainly with glial RNA. In previous experiments differences had been indicated in base amounts in nerve and glial cells of nonstimulated control animals. The glial RNA showed higher A and lower G values than nerve cell RNA (Table 8). Learning animals showed an increase in the A to U ratio (somewhat similar to the change in nerve cells reported in the previous experiment), but no change in base composition resulted for the functional controls (Table 9). A significant decrease in C was observed for the learning animals. Nuclear RNA in nerve cells of a functional control group that was rotated increased about 25 to 30 percent; however, no change in base composition resulted.

Hydén and Egyhazi also dissected nerve cells from the reticular formation in the lower brain stem and extracted nuclear RNA. The average amount of RNA per nerve cell during a four day learning period was: Day 1, 515 $\mu\mu$g; Day 2, 568 $\mu\mu$g; Day 3, 568 $\mu\mu$g; and Day 4, 590 $\mu\mu$g. There were, however, no significant differences in base composition for control and learning animals (Table 10).

Another experiment involving the forcing of right-handed rats to use

TABLE 9 Mean Percent Base Composition of Glial RNA for Experimental and Control Rats (Hydén and Egyhazi, 1963)

Bases	Controls	Functional Controls	Experimentals
Adenine	25.3	25.1	28.3
Guanine	29.0	28.6	28.8
Cytosine	26.5	27.4	24.3
Uracil	19.2	18.9	18.6

TABLE 10 Mean Percent Base Composition of Neural RNA from Reticular Formation (Hydén and Egyhazi, 1963)

Bases	Controls	Learning
Adenine	23.9	22.8
Guanine	25.3	26.2
Cytosine	28.9	29.7
Uracil	21.9	21.3

the left hand to obtain food was conducted by Hydén and Egyhazi (1964). The RNA of neurons from layers 5 and 6 of a small sector of the brain, anterior dorsal cortex, was obtained. This tissue is assumed to contribute to motor behavior in rats. Tissue was obtained both from the left hemisphere controlling the right hand (control) and right hemisphere involved in the learning task (experimental). This procedure allowed for each animal to be his own control. A significant difference occurred, 22 $\mu\mu$g/cell for controls compared to 27 $\mu\mu$g/cell for the other side. Changes in base ratios also resulted (Table 11) with A, G, and U increasing, and C decreasing. In the previous experiments Hydén reported increases in the A/U ratio for learning animals. In the handedness experiment, this ratio was approximately one in both control and experimental tissues.

TABLE 11 Mean Percent RNA Base Composition of Cortical Neurons of Left Side (Control) and Right Side (Experimental) (Hydén and Egyhazi, 1964)

Base	Controls	Experimentals
Adenine	18.4	20.1
Guanine	26.5	28.7
Cytosine	36.8	31.5
Uracil	18.3	19.6

Three right-handed animals not involved in the transfer of hand experiments were analyzed with respect to content and composition of RNA from the same neurons as analyzed in the transfer experiment. No significant difference between the tissue from the right and left hemispheres was found. In another experiment three right-handed rats were allowed to perform the same number of reaches and for the same time as the animals in the learning experiment. A slight increase of RNA in the left side neurons occurred but no base changes resulted.

In a later paper, Hydén and Lange (1965) analyzed data in the wire balancing and hand forcing experiments to determine the change in RNA (ΔRNA) from one period to the next during the learning process. Base analyses indicated that during the early portion (three to five days) of the task, A and U predominated in the ΔRNA. Later, increased amounts of G and C were present. Such results provide the basis for the Hydén selective model of learning discussed in Chapter 11.

The experimentation by Hydén represents the first research in this area. The results, and interpretation of these results, however, are not clear. Each of the learning tasks involves a type of motor activity. One would like to see some research by Hydén utilizing other tasks, some which minimize motor activity. Furthermore, Hydén has tended to analyze base amounts mainly in the brain area concerned with the motor activity involved in the task. Analyses of other brain sites, and nonneural control sites, should provide more complete information.

A word of caution is required concerning the base analyses. Because amounts of each base are expressed as a percentage of the total amount of all bases, a change in one will tend to make for the appearance of change in one or more of the other bases as well. For example, assume that in a single cell the amounts of each base (in $\mu\mu$g) are as follows: A–20, G–30, U–20, C–30. The percentage of each base would be 20, 30, 20, and 30, respectively. Now if there is an increase of 5 $\mu\mu$g in A but the others do not change, the percentages would be 23.8, 28.6, 19.0, and 28.6. These results suggest wrongly that there are changes in all bases. The overall effect of base analyses conducted in this fashion could be to magnify artificially the amount of differences which occur.

The most important aspect of the Hydén research is the base changes. The meaning of these changes, however, is not clear because of the grossness of this measure. Two major alternatives are possible. A qualitative change may be occurring with polymer RNA being modified to produce a new molecular species, new RNA species may be synthesized, or RNA from certain cells may permeate other cells such that the RNA population contains new types of RNA. On the other hand, quantitative changes may be involved, i.e., the relative amounts of the synthesis and/or degradation of species of transfer, ribosomal, and messenger RNA's may be changing. The latter appears to be the most likely interpretation. For example, let us assume that a cell contains only two species of RNA. Assume also that the first contains 20 percent of A whereas the second has 25 percent and that the two are present in equal amounts. A base analysis on this cell would indicate 22.5 percent of A, an averaging of the two species. If during learning no other species of RNA are synthesized, but the first species is 10 times as abundant as is the second, the base analysis would indicate that 20.5 percent of A was present. The difference of 2 percent might suggest that new species of RNA were synthesized during learning, whereas in actuality a statistical artifact is occurring because of the pooling of the two types of RNA. The problem is more complicated in actual cells because hundreds of RNA's are present, each with different amounts of the bases.

There seems to be the implication in the Hydén research that specific types of base changes occur during learning tasks and within cells involved in the learning event. But his research does not clearly show a consistent change. For example, the A/U ratio increased in some cases but not in others. Furthermore, there are a number of studies which indicate base changes in many nonlearning tasks. Egyhazi and Hydén (1961) noted that administration of malononitrile brought about a decrease of 6 percent in C in nerve cells while glial cells showed a 20 percent increase in C and a 25 percent decrease in G. Geiger (1957) reported that electrical stimulation of the cerebral cortex for 30 seconds caused an increase in C and A, whereas the amounts of U and G remained constant. Grampp and Edstrom (1963) found that a six hour excitation of certain cells resulted in an increase in the A/U ratio. Edstrom (1964) showed that marked changes occur in the A/G ratio in nerve fibers of the goldfish after transection of the spinal cord.

The most convincing evidence that a specific base change is not unique to learning is the report by Hydén et al. (1969) and John (1967) in which the base changes which appeared in planarians subjected to a conditioning procedure were the same as those receiving pseudoconditioning (i.e., random pairing of the conditioned and unconditioned stimuli) and other control groups in one type of statistical analysis. With another statistical analysis, differences in base amounts did occur but were in an opposite direction than would be predicted, viz., a pseudoconditioning group showed greater A/U and A/C ratios than did naïve and conditioned planarians. The last two groups did not differ in these ratios. The training of the planarians was achieved in John's laboratory and the biochemical analyses were performed by Hydén.

Table 12 indicates the base amounts for five groups in the planarian experiment. Base analyses were conducted with head and tail portions of the worms. Although the behavioral responses clearly indicated learning only in the experimental group, the base ratio analyses showed either no differences between these animals and four control groups or that base ratios were less for the learning group than for some of the control groups.

A number of other studies have been performed using the direct ap-

TABLE 12 Mean Percent Base Amounts in Planarian Experiment
(Hydén et al., 1969)

Group*		A	G	C	U	A/G	A/C	A/U
Experimental	Head	31.6	19.1	20.3	29.0	1.69	1.57	1.09
	Tail	30.4	20.1	19.7	29.8	1.52	1.55	1.02
Control 1	Head	33.1	15.5	21.4	30.0	2.40	1.56	1.11
	Tail	31.9	16.4	22.6	29.2	1.95	1.41	1.09
Control 2	Head	36.3	19.1	17.8	26.8	1.93	2.05	1.36
Control 3	Head	30.4	19.3	21.3	29.1	1.65	1.43	1.05
	Tail	32.2	17.0	19.1	31.6	1.89	1.69	1.02
Control 4	Head	35.1	13.3	21.0	30.6	2.67	1.68	1.16
	Tail	31.7	16.4	22.0	29.9	1.92	1.54	1.06

Reprinted, with permission, from Hydén et al. (1969). Copyright 1969, Pergamon Press.

* The Experimental Group was presented with paired light and cathodal shock; Control 1, paired light and anodal shock, which does not produce conditioning; Control 2, randomized light and cathodal shock; Control 3 randomized light and anodal-cathodal shock; Control 4, naïve worms.

proach. Shashoua (1968, 1970) performed base analyses on brain RNA from goldfish who were involved in the acquisition of new swimming skills. A foam polystyrene float was attached to the anterior ventral surface of each goldfish, which forced each animal to learn new patterns of swimming so as to maintain himself in an upright position. During the acquisition of the new swimming pattern, Shashoua reported that increases in the U/C ratio resulted. Unfortunately, these results are inconclusive because: (a) the control and experimental conditions confounded learning, motor activity, and stress and (b) the U/C ratio was based only on counts per minute (cpm) and, thus, did not control for the possibility that varying amounts of U and C in each animal and between experimental and control fish might have been responsible for the reported increases.

Dellweg, Gerner, and Wacker (1968) performed a learning experiment similar to the one by Hydén and Egyhazi (1962); rats attempted to go from one compartment to another containing food by balancing on a rope at a slope of about 40°. They found no difference in trained and untrained rats in the amount of tRNA in whole brain, but increases in rRNA and polysome content occurred. The ratio of polysomes to ribosomes in the brains of trained rats increased also. Inasmuch as the presence of polysomes is required during protein synthesis, this latter result suggests that protein synthesis would be enhanced in the trained rats.

In an interesting set of experiments involving 15 minute shock avoidance training with mice performed at the University of North Carolina, a number of results have been found which are pertinent to the involvement of RNA in learning (Zemp et al., 1966; Zemp, Wilson, and Glassman, 1967; Adair, Wilson, and Glassman, 1968; Adair et al., 1968). These experiments showed that more radioactive precursor was incorporated into the RNA extracted from brain nuclei, in brain ribosomes, and in polysomes of the trained animals. By appropriate controls they were able to indicate that the increased incorporation was not due to the effect of light, buzzer, shock, or handling which the mice received. To determine if the label was being incorporated into species of RNA which were qualitatively different from those present during control conditions, the authors sedimented labelled nuclear and rRNA in sucrose gradient centrifugation. The radioactivity in the RNA of the trained mouse was greater than that for the untrained mouse, but the gross pattern was the same for both; furthermore, the sedimentation pattern for the trained animals resembled those found after RNA synthesis had been stimulated by hydrocortisone in the liver or by estrogen in the uterus. These sucrose gradient results are indicators of the molecular weight of RNA which cannot show qualitative differences directly, however; unique RNA species in learning animals may be of the same molecular weight as RNA species in nonlearning animals but could contain different base sequences.

Zemp, Wilson, and Glassman (1967) found that the increased incorporation of uridine into RNA took place in the upper brain stem and associated structures; a small decrease of uridine incorporation took place in the cortex.

Other experiments by Adair, Wilson, and Glassman (1968) were concerned with various behaviors involving the jump box of the shock avoidance task. In each case when the mouse learned to jump to the shelf in response to the conditioned stimulus, increased incorporation of radioactive uridine into brain polysomes occurred. If the behavioral experience did not include avoidance learning, the increased incorporation did not take place. Thus the series of shock avoidance experiments tend to show a relationship between the incorporation of labelled precursors in RNA and this type of learning. It is not clear from these experiments, however, whether the RNA being synthesized includes RNA for this learning task which is qualitatively different from RNA synthesized during other behavioral tasks.

Other experimenters have reported RNA changes during learning. Nasello and Izquierdo (1969) found that shock avoidance conditioning resulted in an increase in the RNA concentration of the parietal cortex (posterior part of brain) and in a subcortical site (dorsal hippocampus) of rats. In the brain of rats, RNA and protein levels and synthesis increased during one-way active avoidance conditioning when the shock level was low, but decreased with a higher shock level (Gaito, Mottin, and Davison, 1968).

Another shock avoidance study with rats reported that increased incorporation of radioactive leucine occurred in the nuclei of hippocampus, entorhinal cortex (ventral portion of posterior cortex), and septal area (a subcortical site), but not in other brain areas or in the liver (Beach et al., 1969).

Bowman and Strobel (1969) trained rats in a Y maze to discriminate spatial cues for water. On reversal training these animals evidenced a 25 per cent increase in incorporation of labelled precursors in hippocampal RNA. Practice of reversals for 60 minutes abolished this increment and produced a decrease in RNA synthesis in basal ganglia (subcortical sites).

An interesting experiment with protozoa (Applewhite and Gardner, 1968) found that greater incorporation of radioactive uridine occurred within one minute in habituating subjects, but the differences in incorporation between these subjects and controls had disappeared by 150 seconds, although the habituation effect was at its greatest at this time. By four minutes, differences in habituation and in incorporation were no longer present.

Although the experimental results by Hydén and others discussed in this chapter are important ones, these results do not clearly show the exact relationship between RNA and learning events. Whether the RNA changes are primary effects, secondary effects, or parallel effects is not certain. One can say only that RNA changes occur during learning.

Chapter 7

RESEARCH CONCERNED WITH THE
PRODUCTS OF DNA ACTIVITY:
HYBRIDIZATION PROCEDURES

One important assumption by Hydén and others who assign a unique role to RNA in learning events is that during these events RNA is synthesized which may be qualitatively different (in primary structure) from RNA which is synthesized during other behaviors. For example, Hydén specifies that an RNA rich in A and U is synthesized during the early stages of learning; later the RNA synthesized has a predominance of G and C (see Chapter 6). The results in the previous chapter, and those discussed in later chapters, do not show clearly that qualitatively different RNA's occur, however. Unfortunately, the procedures utilized have been too gross to detect qualitative changes in primary, secondary, tertiary, or quaternary structures of RNA.

One means of attempting to determine if qualitatively different RNA occurs in learning events is by DNA-RNA hybridization procedures (Gillespie and Spiegelman, 1965; Bonner, 1966; Gaito, 1966). If one heats a solution of rat brain DNA at 95°C for 10 minutes, the double stranded DNA will split into single strands. If this DNA is then poured onto nitro-cellulose membranes, these membranes will "trap" single strands but will allow any double strands to pass through. If a membrane with attached DNA is placed in a solution of RNA, those RNA molecules which are complementary in base sequence to DNA sites will become firmly attached and be resistant to RNase treatment. If this DNA-RNA hybrid is put in another solution of the same RNA, no further hybridization will occur because all DNA sites complementary to the RNA already are occupied. On the other hand, if this hybrid is added to a different solution of RNA which is complementary to other DNA sites, further hybridization will occur.

Putting this procedure within a behavioral framework, the rationale is the following. If there exist unique, qualitatively different species of brain RNA which are synthesized during learning, and RNA from the brain of a nonlearning animal is hybridized with single strand DNA, then when RNA from the brain of a learning animal is added to this hybrid, the unique RNA species should adhere to the DNA. An important aspect of this *successive competition hybridization procedure* is that only the RNA from learning animals is labeled. Therefore, the presence of label in the twice hybridized DNA will suggest that RNA species not present in the brain of nonlearning animals have been synthesized in learning animals during the task.

At the present time a number of behavioral studies are underway in the Molecular Psychobiology Laboratory at York University in which double hybridization procedures, supplemented by appropriate single hybrids, are involved. In all experiments, DNA is extracted by repeated cold phenol treatments and purified. RNA is extracted by cold and hot phenol, and then purified. The hybridization procedures are those of Gillespie and Spiegelman (1965). DNA trapped on membranes is incubated with an RNA sample at 66°C for 12 hours, washed, and treated with RNase. For double hybrids, this procedure is repeated.

Prior to the initiation of behavioral experiments, a preliminary experiment with gastrointestinal DNA indicated that all, or almost all, of the DNA adhered to nitrocellulose membranes. Other preliminary experiments indicated that the amount of RNA hybridizing with 50 μg of DNA seemed to reach a peak with an input of 50 μg of RNA and that hybridization for 24 hours gave the same results as at 12 hours. Thus most experiments used 50 μg of RNA hybridizing with 50 μg of DNA for 12 hours.

Rapid progress has occurred with a shock avoidance conditioning task. In Experiment 1 (Machlus and Gaito, 1968a) labelled RNA from avoidance conditioned rats competed with unlabelled RNA from nonbehaving control animals. Label appeared consistently in the double hybrid, suggesting that the brains of learning animals contained RNA species qualitatively different from those in the brains of nonlearning rats. Results with single hybrids for the two groups of rats, and with double hybrids in which the RNA from nonlearning rats was labelled and the RNA from learning animals was unlabelled, provided support for this conclusion.

In a second experiment (Machlus and Gaito, 1969b) the nonbehaving group of rats was replaced by rats subjected to forced motor activity on a tread mill. Results occurred which were similar to those in Experiment 1, suggesting that the RNA species detected in Experiment 1 probably were not due to the motor aspects of the avoidance task. These results were not conclusive, however, because the motor activities in the two tasks were different.

Experiments 3 and 4 provided more adequate controls than did the previous experiments (Machlus and Gaito, 1969). A shock avoidance rat was injected intracranially with 1 mc (millicurie) of uridine-5-H[3] in 100 microliters of physiological saline. (Uridine goes mainly into RNA as uridylic acid.) There were two shock yoke control animals in each litter; one was injected with uridine-5-H³ and the other, with unlabelled uridine. Ninety minutes later the three littermates were placed in the shock chamber of a one-way active shock avoidance apparatus. The shock chamber was divided into three parts so that each of the rats was isolated. The shock avoidance rat was able to respond to the conditioned stimulus by running from the chamber; the shock rats could not leave this chamber. After 15 minutes of adaptation in the shock chamber of the apparatus, the shock avoidance animal was given 15 trials in 15 minutes and sacrificed by immersion in liquid nitrogen. All shock avoidance animals showed 10 or more avoidance responses in 15 trials. The shock rats received shock when the avoidance rat was shocked; these controls did not receive training and were sacrificed at the end of 30 minutes in the shock chamber, at the same time as the avoidance rat.

Labelled RNA was extracted from the whole brain of the shock avoidance rat and from one shock rat. Unlabelled RNA was extracted from one-half of the brain of the rat that was injected with unlabelled uridine; DNA was extracted from the other half of the brain. This DNA was used for all hybrids because preliminary work indicated that hybridization results with DNA from littermates were similar to those in which DNA and RNA were from the same animal.

With the three rats in each litter, four hybrids were obtained as shown in Table 13. Hybrids 3 and 4 were the crucial ones involving successive competition hybrids; Hybrids 1 and 2 were single hybrids which were utilized as a check on the results with Hybrids 3 and 4. Experiments 3 and 4 differed only in one respect: in Experiment 3, 100 μg of RNA was hybridized with 50 μg of DNA; 500 μg of RNA was annealed to 20 μg of

TABLE 13 Estimated Amounts of RNase Resistant RNA in Single and Double Hybrids

Hybrids	Exp. 3 Mean	S.D.	Exp. 4 Mean	S.D.	Exp. 5 Mean	S.D.	Exp. 6 Mean	S.D.
1. DNA-RNA$_{SA}$*	1.40	.173	0.62	.100	2.68	.374		
2. DNA-RNA$_S$*	1.08	.114	0.44	.084	1.80	.281		
3. DNA-RNA$_S$-RNA$_{SA}$*	0.64	.153	0.27	.075	0.76	.200	0.00	0.00
4. DNA-RNA$_S$-RNA$_S$*	0.02	.019	0.00	.000	0.00	0.00	0.00	0.00
5. DNA-RNA$_{SA}$-RNA$_{SA}$*					0.00	0.00		
6. DNA-RNA$_{SA}$-RNA$_S$*					0.00	0.00		

RNA$_{SA}$, RNA from shock avoidance animal; RNA$_S$, RNA from nonlearning shocked animal; asterisk indicates the presence of labelled precursor in RNA; S.D. is standard deviation. DNA and RNA were from brain in Experiments 3, 4, and 5; in Experiment 6, from liver. There were 12 observations for each hybrid in Experiments 3 and 5, 8 in Experiment 4, and 16 in Experiment 6. In Experiments 3, 5, and 6, 50 μg of DNA was hybridized with 100 μg of RNA; in Experiment 4, 20 μg of DNA, with 500 μg of RNA.

DNA in Experiment 4 to provide a greater RNA/DNA ratio. This ratio in Experiment 3 was 2; in Experiment 4, 25.

In all experiments, the counts per minute (cpm) were obtained for each hybrid and converted to disintegrations per minute (dpm) using a quench correction curve. The specific activity of RNA* for each littermate was determined from a 50 μg sample in which the cpm were converted to dpm. The specific activity in the various experiments varied between approximately 1000 and 2500 cpm/μg RNA. There appeared to be no difference in specific activities between shock avoidance and shock rats. The amount of labelled RNA hybridizing in each hybrid was obtained by comparison with the specific activity of the RNA for the rat from which the hybridized RNA was obtained.

The results of Experiments 3 to 6 are presented in Table 13 as amounts of RNase resistant RNA present in each hybrid. Two samples were obtained for each of six litters in Experiment 3 (12 in all), but only one sample was possible for each of eight litters in Experiment 4. The results in both experiments were clear. In all samples (12 in Experiment 3; 8 in Experiment 4), Hybrid 3 showed values much greater than zero. Hybrid 3 was greater than Hybrid 4, and Hybrid 4 was zero or near zero. Likewise, in all samples, the amount of RNA hybridized for the shock avoidance rat (Hybrid 1) was greater than that for the shocked rat (Hybrid 2). The amounts hybridized with 20 μg of DNA in Experiment 4 were proportionately the same as those achieved in Experiment 3 with 50 μg of DNA. The single and double hybrids consistently suggested that qualitatively different species of RNA were present in the brains of shock avoidance learning rats than were present in the littermate shocked animals. Thus the results of the four experiments consistently suggested the presence of unique brain RNA, using RNA/DNA ratios of 1, 2, and 25.

Another study, Experiment 5, was conducted with further controls to provide double hybrids for all possible sequences of shock avoidance and shock conditions. A second shock avoidance rat (injected with unlabelled uridine) was added. With the four rats in each litter, six hybrids were obtained (Table 13). Hybrids 3 to 6 were the crucial ones involving successive competition hybridization. Six litters of four rats were used. Two samples were obtained for each litter in Experiment 5 (12 in all). In all 12 samples Hybrid 3 showed values of μg RNA much greater than zero whereas the other double hybrids were at background levels. Likewise, in single hybrids, for all 12 cases, the amount of RNA hybridized for the shock avoidance rat was greater than that for the shocked rat. With cpm or dpm as the dependent variable, the results were the same. These results are consistent with those of the previous experiments and suggest the presence of unique RNA species during shock avoidance conditioning.

* The cpm per μg of RNA.

Another experiment (Experiment 6) was undertaken to determine if these unique species were present in the liver during conditioning. Four litters of three rats each were used. Two rats were injected with labelled uridine (shock avoidance trained and shock control) and the other, with unlabelled uridine (shock control). Only Hybrids 3 and 4 of Table 13 were used (DNA-RNA$_S$-RNA$_{SA*}$ and DNA-RNA$_S$-RNA$_{S*}$). Fifty μg samples of RNA from shock avoidance animals indicated the synthesis of labelled RNA in the liver; however, no label was detected in either hybrid for 16 samples (four observations for each of the four litters).

Experiment 7 was conducted to investigate the possibility that the unique RNA species were localized in specific brain areas. Three pairs of littermate rats were used. One of each pair received labelled uridine; the other, unlabelled uridine. Both rats were trained to avoid shock. The brain of each rat was dissected into three parts: cerebral hemispheres, brain stem, and cerebellum. Three single hybrids were obtained, one for each of the three parts. Six double hybrids were used, one each for the six possible sequences of the three parts in successive competition:

$$RNA_{CH}\text{-}RNA_{CB*}, \qquad RNA_{CH}\text{-}RNA_{BS*},$$
$$RNA_{CB}\text{-}RNA_{CH*}, \qquad RNA_{CB}\text{-}RNA_{BS*},$$
$$RNA_{BS}\text{-}RNA_{CH*}, \qquad RNA_{BS}\text{-}RNA_{CB*}.$$

Three observations were obtained for each. The amount of RNase resistant RNA hybridizing for the three brain parts were: cerebral hemispheres, 2.43 μg; brain stem, 2.37 μg; cerebellum, 2.46 μg. No double hybrid showed greater than background label.

Because *in vivo* labelling of RNA is not of a uniform nature, the estimation of the amounts of RNA hybridized by liquid scintillation spectrometry must be considered as an approximation. However, the results have been consistent in this series of studies. The differences have been substantial in indicating the presence of unique RNA, and this conclusion occurs whether one uses cpm, dpm, or amount of RNA as the dependent variable of concern.

The percent of DNA sites occupied by RNA from learning and non-learning animals can be estimated from these results by the following:

$$\text{Percent DNA sites occupied} = \frac{\mu\text{g RNA in single hybrid}}{\mu\text{g DNA used}} \times 100$$

The percentages for the learning animals in Experiments 3, 4, and 5 are 2.8, 3.1, and 5.4, respectively. The shock animals show 2.1, 2.2, and 3.6 percent, respectively. These percentages are approximately what one would expect: about 3 to 5 percent of DNA sites concerned with brain functioning. Presumably, other tissues would show about the same values for specific functions.

The results of these experiments suggest that during this behavioral task, RNA species were produced throughout the three brain portions (CH, CB, BS) which were qualitatively different from those present in the brains of nonlearning rats and were not present in the liver. During the first 105 minutes of the incorporation period, the learning and nonlearning animals were treated the same. Thus the differences probably reflect the synthesis of RNA during the last 15 minutes during which learning was occurring. Other individuals have reported what appears to be an increase in the rate of synthesis of RNA in the brains of mice after 15 minutes of training in a shock avoidance task (see Chapter 6).

Thus the results of this series of studies suggest the synthesis of unique, qualitatively different species of RNA during this learning task, a conclusion which is consistent with those of Hydén and others who hypothesize that RNA has a unique role in learning events. Other experiments, however, are being conducted to evaluate the possibility that these results are artifactual in nature, because the hybridization techniques are very difficult ones and have serious pitfalls.

Experiments are underway with other behavioral tasks (discrimination learning, visual stimulation*) and further experiments are anticipated using the shock avoidance task. These experiments will employ hybridization procedures which will provide comparisons between control and experimental rats over varying, including large, amounts of RNA. Other experiments are attempting to determine the physical and chemical characteristics of this RNA.

An interesting aspect of these results is the concern as to what these unique RNA's are. It is too early to suggest the exact role that these RNA's play in learning (assuming, of course, that they are not artifactual). Recently the idea has been advanced that much of the DNA of higher organisms is made up of sequences which recur anywhere from a thousand to a million times per cell (Britten and Kohne, 1968). During DNA-RNA hybridization, these repeated sequences in DNA anneal rapidly with RNA even when the RNA concentration is low. With the short period of incubation (12 hours) and the low concentration of RNA which have been used, it is probable that the unique RNA is annealing to the repeated DNA sequences (J. Bonner, personal communication). The unique RNA species may be mRNA or nuclear RNA which functions to derepress other DNA sites (J. Bonner, personal communication; activator RNA of Britten and Davidson, 1969). If they are nuclear RNA, e.g., a type of activator RNA, they may function

---

* As this volume goes to press, part of the visual stimulation study is complete. Using improved hybridization procedures, unique species were not detectable during visual stimulation, and RNA/DNA ratios of 50 and 75 did not saturate the DNA sites. This lack of saturation is inconsistent with the results of the shock avoidance experiments in which RNA/DNA ratios of 1, 2, and 25 appeared to saturate DNA sites. Further research is underway to reconcile these discrepancies.

to increase the synthesis of RNA from DNA sites which have been continuously active, but at a low level, or they may activate DNA sites which were repressed. In the former case, only the activator RNA's would be unique (i.e., qualitatively different); in the latter event, both the activator RNA and the mRNA from the depressed site would be unique species.

Even if these RNA's are shown consistently to be qualitatively different from the RNA under nonlearning conditions, the relationship between these RNA's and learning is not resolved. Results indicating specific types of RNA present *only* during learning logically suggest that the primary effects relationship holds; however, the hybridization procedures have not eliminated conclusively the secondary and parallel effects possibilities. In the future these procedures, supplemented by other techniques, may be able to obtain results which can differentiate between the three possibilities. For example, one might separate the unique RNA from the other RNA trapped on the membranes and inject the unique RNA (in sufficient quantity) into the brain of naïve animals. If these injections consistently facilitated learning, these results would be strong evidence that RNA has a primary effect in learning.

Chapter 8

RESEARCH CONCERNED POSSIBLY WITH BOTH
DNA REGULATION AND THE PRODUCTS OF
DNA ACTIVITY: ADMINISTRATION OF
RNA OR BRAIN HOMOGENATES

The indirect approach of administrating specific chemicals and observing the behavioral results has been a popular one, possibly because of the relative ease of completing experiments of this nature. With this approach the investigator usually assumes that the chemical produces a single *specific* neurochemical effect, an assumption which is quite hazardous.

Assuming that RNA produces a primary effect on learning, it seems quite logical to administer some of this substance with the expectation that learning should be enhanced by providing the necessary RNA, or that it should stimulate DNA sites to synthesize RNA. Investigators have used either yeast RNA or RNA (or brain homogenates) extracted from trained and untrained subjects.

Yeast RNA

The earliest experiments using yeast RNA were those conducted by Cameron. For example, Cameron and Solyom (1961) found that administration of yeast RNA (but not DNA) to aged individuals brought about memory improvement. These changes involved almost total retention in some cases. When the RNA was discontinued later, memory relapses occurred.

Other memory improvement results were reported by other researchers (Kral and Sved, 1963). They also indicated that the RNase activity in the blood increased with age. They suggested that the yeast RNA administered to the elderly patients in their studies was attacked by the RNase, and this event allowed more of the native RNA of the human to be free from attack.

RNA studies have been conducted with animals also. Cook et al. (1963) injected doses of RNA intraperitoneally (IP) in rats for three days, one week, two weeks, one month, and 53 days. The rats were placed in a chamber with an electrified grid floor, a pole suspended from the top center, and a buzzer. They were given trials consisting of the pairing of the buzzer and shock (through the grid floor). Later only the buzzer was presented. Each trial was terminated by the rat jumping onto the pole after the onset of the buzzer, or at the end of 30 seconds. The authors reported that the RNA groups were superior to control groups given saline in the acquisition of a pole climbing response in all except the three day group; however, resistance to extinction was greater in the RNA groups for all time periods.

Wagner, Gardner, and Beatty (1966) reported an enhanced acquisition effect similar to Cook et al. for rats administered RNA for at least 30 days. Because acquisition was not enhanced for an RNA group in a second experiment with a Y maze, these authors suggested that the effect of RNA was on general activity or in sensitizing the subject rather than on learning processes directly.

In a series of experiments Corson and Enesco (1966) investigated the effects of yeast RNA on a variety of tasks. An RNA group was superior to a control group in acquisition of the pole climbing response but not in other shock motivated tasks requiring discrimination, food rewarded, and water rewarded responses; also no differences appeared in open field activity or in basal metabolic rates. The experiments were performed with the same subjects in succession. Since they obtained positive results only in the shock escape pole climbing situation, and there is evidence that IP injected material does not reach the brain (see below), the authors suggested that the effects of RNA (when present) would be mediated primarily by processes outside the brain.

Other yeast RNA experiments with humans and rats were discussed by Gurowitz (1969). The results of these experiments have not shown consistently an enhancement effect on behavior.

An important concern in these studies using the IP route of administration (or any extra brain route) is whether the intact RNA reaches the brain. Studies show that some intact RNA can penetrate cells, e.g., in peritoneal cells following IP injections (Schwarz and Rieke, 1962), in ascites tumor cells from mice (Galand, Remy, and Ledoux, 1966). However, in these experiments the injected RNA traveled only a short distance before entering the cells of concern. A number of investigators using extra brain injections of labelled RNA in mice and rats have reported that little, if any, label enters the brain (Eist and Seal, 1965; Enesco, 1966; Luttges et al., 1966; Sved, 1965). Both Enesco (1966) and Sved (1965) used intravenous administration of RNA and concluded that the RNA was degraded to free bases and ribose and used in cellular functions. Based on these results which indicate that RNA is degraded and appears not to reach the brain, one could infer that any positive behavioral results might be due to the effect of the degraded RNA constituents on organs other than the brain, e.g., liver. The RNA base constituents provide necessary metabolites for the metabolic pool and thus influence the overall performance of the organism, not affecting learning aspects alone; the ribonucleotides, or portions of the nucleotides, are constituents of numerous cellular products such as ATP, GTP, UTP, CTP, a number of coenzymes, and phospholipids which are essential for normal neural function. For example, A nucleotides play an important role in energy metabolism; C nucleotides participate in the synthesis of lipids; nucleotides of U and G take part in the synthesis of polysaccharides; and G nucleotides function in the synthesis of proteins (Mandel, Harth, and Borkowski, 1961).

Even in studies in which RNA is injected into the brain directly, one would expect that much of the RNA would be degraded. The brain is rich in enzymes which break down RNA. Thus any positive behavioral results in these studies could be attributed to the effects of RNA constituents on general cellular processes in the brain.

The increased resistance to extinction of the RNA groups in the Cook et al. study raises a question as to the effects of the yeast RNA on the adaptability of an organism. In a learning task, acquisition of a specific response is adaptive for the organism. For example, after a few trials a rat acquires a lever pressing response which allows him to obtain food. During extinction (i.e., when food no longer accompanies the press of the lever), the tendency to respond gradually disappears. However in some of the animals administered specific chemicals, the extinction process is quite different than in normal animals, e.g., in the Cook et al. study and others to be discussed below, especially with magnesium pemoline.

These studies tend to suggest a rigidity in responding unlike that of the normal animal. These animals tend to persist in responses long after they are no longer adaptive. For example, Cook et al. (1963) indicated in a diagram that in 30 day treated animals, all rats of the control group extinguished by 24 trials. At this point about 70 percent of the RNA treated rats were still responding. At 41 trials when the experiment was terminated, 20 percent of the experimental rats persisted in the response. Thus one could hypothesize that the effect of the chemical is affecting the activity level (or some general function) of the organism, either in the brain or elsewhere. This effect would facilitate the acquisition of some specific responses which are adaptive but also would tend to predispose the animal to continue blindly with responses when they are no longer of value to the organism.

RNA or Homogenates from Trained Animals

Of the various methods utilized to differentiate between the primary effects, secondary effects, or parallel effects of RNA and protein, the transfer experiment appears logically to be the most appropriate one. If RNA or protein has a primary effect on learning, one should be able to facilitate learning of a specific task in naïve animals by removing these chemicals from animals trained in this task and administering them to the naïve recipients.

Thompson and McConnell (1955) reported that the planarian, a flatworm, was capable of acquiring a conditioned response (contraction to light following a series of trials in which light and shock are paired). When cut in half following conditioning, the worm regenerated head and tail animals which were reported to show memory for the previous conditioning as indicated by requiring fewer trials to reach a criterion in later conditioning training.

Zelman et al. (1963) trained planarians in a conditioning situation. Then RNA was extracted from these worms and from untrained worms. The RNA from trained worms was injected in one group of planarians and RNA from untrained worms, into another group. They found some tendency, but not a consistent one, for the animals fed the first RNA to respond more frequently to the conditioned stimulus than did the animals fed RNA from the untrained worms.

A similar experiment was conducted by Fjerdingstad, Nissen, and Roigaard-Petersen (1965) using rats. They obtained RNA preparations from animals which had been conditioned in a simple maze (Group 1) and from nonconditioned rats (Group 2). Then they injected the RNA from the first group into the brains of a group of naïve recipients. The same volume (but not necessarily the same amount) of RNA from Group 2 (50 μl/animal) was injected in another naïve group. Another group contained uninjected controls. The group injected with RNA from trained animals was significantly superior in maze performance in two test sessions of about 15 reinforcements each and in later trials.

A number of other research teams reported facilitative effects also in a variety of tasks (e.g., Babich et al., 1965; Nissen, Roigaard-Petersen, and Fjerdingstad, 1965; Reinis, 1965; Jacobson et al., 1966). A recent study by Faiszt and Adam (1968) reported that facilitative effects were produced if the recipient rats received rRNA but not with other RNA fractions. Their report ended with the cogent caution that "Our results do not indicate whether they involve the specific transfer of a memory trace or the nonspecific facilitation of conditioning" (p. 368).

There are negative results which contrast with these findings. Workers in seven laboratories attempted to replicate the procedures of the above researchers and obtained no facilitative results in 18 experiments (Byrne et al., 1966). In a set of interesting experiments, Luttges et al. (1966) used a number of behavioral tasks and injections both into the brain and into the peritoneal cavity (where most researchers had administered the RNA solutions). They systematically varied the training and testing methods for both donor and recipient animals. The tasks were: (a) a two-choice brightness discrimination in a water Y maze (trained to the darker side); (b) training to the left alley of a two-alley runway; (c) light-dark discrimination in an automated shuttle box (trained to avoid foot shock); (d) pretraining in a straight alley for water reward; (e) training in a modified Lashley III maze; and (f) shock avoidance. The interval from time of injection to testing session was varied also. In no case was there any facilitative effect of the RNA treatment, even when different concentrations were used (RNA extracted from one half brain to three brains per injection). Phenol extraction procedures were used as in the positive transfer experiments. This series of experiments was executed with rats and mice as subjects. The authors also reported that activity rates (measured by number of crossings in an open field) were the same for all groups.

Similar negative results were obtained by Gross and Carey (1965), Gordon et al. (1966), and Halas et al. (1966). The latter authors reported differences in general activity measures. An experiment in the author's laboratory (Schaeffer, 1967), reported no facilitative behavioral result nor a change in the brain neurochemical pattern when labelled RNA was injected into the brain. Furthermore, Halstead (personal communication) found no facilitative effect consistently.

Branch and Viney (1966) obtained no statistically significant facilitation of position discrimination learning with RNA extracts. The results, however, were in the direction favoring the rats injected with RNA from trained rats. The authors suggested that a facilitation effect might be demonstrated if increased training were allowed. In a later experiment Viney, Branch, and Gill (1967) used increased donor training but found no significant facilitative effect. They reported differences between the performances of injected groups (those receiving RNA from trained animals; those with RNA from untrained rats; and those receiving a saline solution) and the original donor rats which suggested a general activation effect of the IP injections.

One study (Essman and Lehrer, 1966) showed that RNA extracts from either the brain or the liver of donors produced a positive behavioral effect in a simple water maze by the recipient animals. This indication that liver RNA in learning animals is capable of precipitating a transfer effect argues strongly against the hypothesis that memory is being transferred, unless one assumes that memory is recorded in the liver.

Thus the results suggest that the specific effect of RNA from trained animals is too subtle to be detected consistently or that there is no effect on learning. In the former case, it could be argued that the specific effect is so subtle that it was masked by confounding variables in the experiments reporting no facilitative result. In the latter case the positive results may be attributed to Type I statistical errors,* experimenter bias, a general metabolic effect brought about by differential concentrations of RNA injected into the experimental animal (see below), the contributions of other chemicals, or other causes.

* Accepting differences as real ones when differences are due to chance.

The positive effects obtained in the RNA injection experiments might be due to polypeptides or proteins* contaminating the RNA preparation rather than to RNA. The extraction methods of some individuals leave doubt as to the purity of their extractions, protein and DNA being present with RNA (Fjerdingstad, Nissen, and Roigaard-Petersen, 1966). This conclusion that polypeptides or proteins are responsible was reached by Ungar and Oceguerra-Navarro (1965). A whole brain homogenate was prepared from sound habituated rats and injected IP into mice (habituation defined as absence of a startle response to sound). They reported that mice injected with this homogenate took 2.1 days to habituate, while the saline injection control group took 12.0 days. In another experiment (Ungar and Cohen, 1965), tolerance to morphine was reported to have been induced by injections of brain homogenate from morphine addicted dogs and rats into mice. The authors state, after examining the properties of what they considered to be the active portion of the injected homogenate, that the effect of their treatment was due to a polypeptide rather than to RNA. This conclusion was based on chemical tests. When they treated the brain extract with chymotrypsin (a protein degrading agent) prior to injection, no facilitation occurred. If the extract was incubated with RNase, the facilitation effect was still present. Other chemical reactions suggested also that the component responsible for the facilitation was of protein nature.

Rosenblatt, Farrow, and Rhine (1966a, b) obtained positive behavioral results in a series of experiments using different methods of extraction and cellular fractions. Their conclusions were that the "information-bearing molecule" appeared to have properties more consistent with those of a polypeptide adhering to the cell membrane rather than of RNA.

Albert (1966) reported facilitative behavioral results following administration of homogenates from trained animals; however, he attributed the effect to RNA rather than to protein. He suggested that the molecules responsible for the transfer effect were labelled, similar to the "recognition" phenomenon of antibodies, so as to migrate to specific tissues. However, there was no indication in the experiment of Schaeffer (1967) that the injected labelled RNA migrated into areas different from that of the control group. The distribution of radioactively labelled RNA and RNA precursor were the same for all groups.

Other individuals obtaining positive results with the injection of homogenates are Byrne and Samuel (1966) and Dyal, Golub, and Marrone (1967).

Gurowitz (1967) found no facilitation of learning following injections of homogenates. A transient metabolic depression occurred which Gurowitz thought might explain some of the positive findings. He also investigated the effects of brain homogenates from trained animals on food and water

* The difference between polypeptides and proteins is one of degree—proteins are larger molecules than polypeptides but both contain the basic units, peptides.

intake and on general activity by the recipients. He found a transitory depressant effect on food intake and on gross activity level. He suggested that the depressant effect of brain homogenates might have been instrumental in the habituation transfer of Ungar and in other studies showing positive results by making the animal less distractible.

A number of individuals have summarized the results of the transfer experiments in an attempt to resolve the discrepancies which abound, e.g., Gurowitz (1969) and Ungar (1971). Ungar has been firm in his belief that the transfer effect is a real one. Based on an analysis of the experiments with positive results and those showing negative results, he suggested that the following conditions provide the highest probability for the obtaining of positive results in the recipients. He maintained that the experiments with negative results did not fulfill one or more of these conditions.

1. Training of the donor animals should be between 6 and 12 days.
2. Crude homogenates should be used. These can be purified in steps, noting the results on behavior in each case.
3. Large doses of brain extracts are necessary, i.e., more than two brain equivalents.
4. Testing should not be conducted before 24, 48, or 72 hours after the injection.

The experiments utilizing the administration of homogenates have had greater success in producing a facilitative effect than have those which have used RNA extractions. Thus the effect (when it occurs) may be due to RNA, DNA, proteins, or some other molecules, or a combination of these. Although these experiments are interesting ones, the results at this time do not provide conclusive answers to five major questions:

1. Is there a transfer effect?
2. What is the molecular system which is responsible for the effect (if it occurs)?
3. Is this molecular system specific to the brain?
4. What are the molecular mechanisms underlying behavioral specificity, i.e., animals injected with material from dark avoidance trained animals showed dark avoidance tendencies, etc?
5. What is being transferred—memory or facilitation?

There does seem to be increasing evidence suggesting an affirmative answer to Question 1. Those researchers with systematic research programs who use brain homogenates have tended to obtain positive results under specific experimental conditions. On the other hand, many of the negative results occurred in "one shot" experiments which did not meet the conditions specified by Ungar (see above) as necessary for obtaining the transfer effect.

With regard to Questions 2 and 3, an assumption by many of those who use injections into tissue outside the brain (both those who inject RNA and those who administer homogenates) is that intact molecules will pass the blood-brain barrier. This assumption appears to be negated by a number of experiments. Eist and Seal (1965) injected C^{14} labelled yeast RNA intravenously (IV) into rabbits and noted the uptake of label in neuronal and nonneuronal tissue through varying time intervals; they reported that the blood-brain barrier effectively excluded the passage of labelled RNA. These findings seem consonant with those of Luttges et al. (1966) who reported that P^{32} labelled RNA injected IP did not pass the brain-blood barrier, according to radioactivity determinations taken up to 23 hours later.

Sved (1965) reported that IV injections gave very small uptake of label into nucleic acids of brain tissue, with rapid breakdown of RNA to free bases and to ribose sugar. Enesco (1966) also reported that IV injections of C^{14} labelled RNA yielded large counts in the site of the injections but were essentially negative in brain tissue.

Some of the individuals that have obtained positive transfer effects have injected the RNA extract directly into the brain. Much of this RNA would be degraded to nucleotides, however, for nucleases (enzymes which degrade nucleic acids) are present in the brain tissue. Thus, little intact RNA should be available in the brain even with intrabrain injections.

Presumably any proteins or polypeptides that were injected would suffer the same fate as RNA and would be widely distributed throughout the body with little, if any, reaching the brain; if injected directly into the brain, most of these materials would be degraded to peptides.

The research results are quite inconclusive relative to Questions 4 and 5.

One control that appears to be lacking in most studies with RNA or homogenates is that of ensuring that *equal amounts* of RNA, proteins, or other presumed important chemicals are injected into the recipients. Most studies do not estimate the amounts which are extracted from donors, in order to equalize amounts injected into recipients. Much research (e.g., see Gaito, 1966) indicates that during stimulation, motor activity, and learning events, increases occur in RNA, proteins, and lipids. If we assume that there is approximately 1000 μg of RNA in the brain of the normal nonbehaving control rat, and that an increment of 20 percent results during behavior, the brain of the experimental rat would contain 1200 μg of RNA. If two or three brain equivalents are used (as Ungar suggests), the brain extracts from the trained animals which are to be injected will contain 400 or 600 μg more RNA than will the brain extracts taken from control rats. As one increases the number of brains used for a single injection, the greater will be the discrepancy in amounts of RNA. Likewise, the same argument would hold for proteins, lipids, or any chemical. Thus the positive results in some experiments may be due to differential amounts of these or other chemicals. The fact that facilitation of performance sometimes occurs with the administration of yeast RNA is also consistent with this interpretation.

Based on the enhancement of behavior which sometimes occurs with yeast RNA, the positive transfer effect resulting with liver RNA, the possibility that greater amounts of chemicals are available per brain for behaving animals, the results which indicate that little of the material injected via an extra brain route actually reaches the brain, and that the constituents of degraded RNA and protein function in many cellular metabolic events, it appears highly unlikely that the transfer effect is specific to the brain or is a memory transfer effect.

Thus, although the transfer experiment appeared to be an excellent possibility for differentiating between primary, secondary, and parallel effects, it has failed to do so in conclusive fashion at this time.

One major step which researchers should take in these experiments is that of determining brain biochemistry in the recipient animals along with behavioral data, i.e., combine the indirect and direct experimental approaches. It would be desirable in these studies to include biochemical analyses of labelled polymer RNA and of nucleotides, nucleosides, and bases in the cell pool; information concerning labelled proteins and precursors in the cell pool could be obtained also. These results could then be related to the behavioral modifications for a clearer picture of cellular events.

Chapter 9

RESEARCH CONCERNED POSSIBLY WITH BOTH DNA REGULATION AND THE PRODUCTS OF DNA ACTIVITY: ADMINISTRATION OF OTHER CHEMICALS

Within the indirect approach, another alternative to the administration of RNA is the giving of chemicals which affect the information flow from DNA through RNA to proteins. These chemicals could either affect the synthesis or degradation of RNA and protein.

Administration of Chemicals which Affect RNA Synthesis or Intact RNA

RIBONUCLEASE

Corning and John (1961) conditioned a number of planarians and then transected them into head and tail sections. They hypothesized that RNA might play a role in the transmission of an acquired structural configuration from the trained to the regenerating tissues. Thus they reasoned that if the trained portions were regenerated in the presence of RNase, the enzyme would affect the altered RNA structure producing some animals with a naïve head and trained tails and others with trained heads and naïve tails. They stated that the head region would probably be dominant; thus the trained head animals would show more retention. They reported that heads regenerated in RNase retained the memory as well as did head and tail sections regenerating in pond water, but the tails regenerating in RNase performed randomly. The authors suggested that the RNase did not affect intact tissue but did interfere with regenerating tissue, and they maintained that the results are compatible with the assumption that RNA is involved in memory events.

This experiment has not been repeated by Corning and John or by other investigators. Thus although it is an interesting experiment, the conclusions should be viewed cautiously.

8-AZAGUANINE

Dingman and Sporn (1961) performed two experiments with 8-azaguanine injections in rats. Eight-azaguanine was used as an inhibitor of RNA because this base analog had been shown to be an inhibitor of enzyme synthesis in bacteria. In both experiments appropriate procedures indicated that the base analog had been incorporated into the RNA of the brain. In neither experiment was there a significant difference between animals injected with 8-azaguanine and control rats in average time to run the maze, suggesting that 8-azaguanine had no adverse effect on the motor ability of the animals. In one experiment the 8-azaguanine animals had a significantly greater mean number of errors than did the controls on all 15 trials in the learning of a maze. In another experiment concerned with retention of a maze pattern (tested by single trial after learning a maze), experimental animals did not differ significantly from control animals, even though the former had a greater average number of errors than did the latter. There were only 8 animals used in each group (as compared with 14 and 15 in the learning experiment); thus if n had been larger in the retention experiment the power of the statistical test would have been greater and the results might have indicated that 8-azaguanine adversely affected both learning and retention of maze pattern in rats.

Based on these results, Dingman and Sporn maintained that RNA may be directly involved in learning (primary effects) but not in retention. However, they admitted that their results did not necessarily indicate that RNA metabolism was intimately linked with the formation of memory traces in the brain because 8-azaguanine might have interfered with metabolic processes which affected RNA indirectly (parallel effects).

Gerard (1963) and Chamberlain, Rothschild, and Gerard (1963) reported that 8-azaguanine had no effect on either avoidance conditioning or maze behavior.

ACTINOMYCIN

Many investigators believe that RNA synthesis is necessary for learning events to occur (a primary effects interpretation). One way to test this possibility is by the use of actinomycin-D, an antibiotic which binds G sites on DNA molecules and interferes with the synthesis of RNA (Goldberg, Rabinowitz, and Reich, 1962). In an interesting experiment, Barondes and Jarvik (1964) found that mice were able to learn a shock avoidance task even though RNA synthesis was inhibited 83 percent throughout the brain by actinomycin-D. In later work, Barondes and Cohen (1966a) reported that animals with 94–96 percent inhibition of cerebral RNA synthesis learned Y and T mazes as well as control animals and showed memory for these tasks as well as controls up to four hours later. Landauer and Eldridge (1966) repeated the experiment by Barondes and Jarvik with similar results. Appel (1965) also found that actinomycin-D had no effect on acquisition

of some tasks. Other research, however, does show a deleterious effect sometimes on learning (Appel, 1965; Meyerson, Kruglikov, and Kolomeitseva,
1965). Agranoff et al. (1967) found that this antibiotic impaired retention
in goldfish if injected up to one hour after training but had no effect if
three hours elapsed between training and injection. To complicate the
picture, Batkin et al. (1965) found that actinomycin-D facilitated the
learning of a T maze by carp and Goldsmith (1967) reported impaired
performance on a passive avoidance task for some rats but not for others.
Such inconsistent results may occur because of the toxicity of actinomycin
and its possible effects on general functions of the cell; the dosage used is
of lethal nature such that the animals tend to die within a few days.

An analysis of the functioning of nerve cells following exposure to
actinomycin-D should be pertinent to this problem. Edstrom and Grampp
(1965) found that RNA synthesis could be inhibited almost 100 percent
for 24 hours without affecting the impulse generating and firing capacity
of the slowly adapting nerve cell of the stretch receptor organ of the lobster.
During this period a fraction amounting to 10–20 percent of the total RNA
was lost from the cell body. Such results show that RNA synthesis is not
necessary for functioning of this nerve cell during this period and suggest
that RNA synthesis may not be necessary in brain cells for learning situations. Any deleterious effects on learning could result because the RNA
changes are of parallel effects nature and affect general cellular metabolism.

TRICYANOAMINOPROPENE (MALONONITRILE)

Egyhazi and Hydén (1961) found that administration of malononitrile
to individuals with certain psychic disorders increased the content of RNA
and proteins in cells of the central nervous system. They indicated that the
malononitrile action was due to the formation of a dimer of malononitrile,
tricyanoaminopropene (TCP). This chemical is presumed to hasten RNA
synthesis; it has an antithyroid effect but causes no observable toxic effect if
given in suitable amounts. They reported that small amounts of this compound caused an increase of 25 percent in the amounts of proteins and
RNA in nerve cells and a decrease of 45 percent in glial RNA. The C in
nerve cells decreased significantly and in glial cells the C showed a 20 percent
decrease. The G in glial RNA also decreased by 25 percent.

Jacob and Sirlin (1964) reported that this drug greatly stimulated the
incorporation of uridine in RNA of insect salivary glands. The drug also
was able to prevent actinomycin-D from inhibiting RNA synthesis.

Gerard (1963; Chamberlain, Halick, and Gerard, 1963) indicated that
malononitrile will increase the fixation of a postural asymmetry. A unilateral
lesion in the cerebellum produces a postural asymmetry in the legs which
is due to an asymmetrical relay of impulses coming down the two sides of
the spinal cord. By cutting the cord the assymmetry may be abolished. They
found that if the cord was cut within 45 minutes the asymmetry was usually

eliminated; otherwise, the asymmetry persisted indefinitely. By speeding up RNA synthesis in neurons with malononitrile on each of four days prior to lesioning, the fixation time for asymmetry was decreased to 25 to 30 minutes. Using 8-azaguanine to retard RNA synthesis, the fixation time was increased to 70 minutes. Chamberlain, Halick, and Gerard (1963) also reported that a single injection of malononitrile in rats prior to an avoidance task did not facilitate acquisition on the day of injection; retention, however, was greater 24 and 48 hours later for these animals than for control rats; in a further study with maze conditioning no facilitation in performance occurred with injections of malononitrile over a period of days.

Many investigators have found that electroconvulsive shock (ECS) will reduce memory for a learning task if given shortly after the task is completed. Essman (1966) reported that if mice were pretreated with TCP for three days prior to a conditioning trial, they showed a significantly high degree of retention in spite of the ECS. The animals treated with TCP also had a significant elevation of RNA levels in brain tissue.

Brush, Davenport, and Polidora (1966) did not find a facilitative effect in the learning and retention of avoidance and water maze tasks following a single injection of TCP prior to conditioning. Solyom and Gallay (1966) reported that TCP in aged rats acted as a stimulant and affected performance, rather than acting directly on learning processes.

Gurowitz, Gross, and George (1968) found that a single injection of TCP did not facilitate, but interfered with, passive avoidance learning. They maintained that the short-term effect was a stimulant one which interfered with learning and that chronic administration of the drug was required to prevent this effect. Gurowitz (personal communication) maintained that short-term administration of TCP causes a stimulant effect which interferes with discrimination learning but that longer term administration facilitates this learning. There appears to be some evidence in the literature for this possibility with TCP or other chemicals, e.g., yeast RNA, as many of the studies reporting a facilitative effect on learning have used chronic administration.

MAGNESIUM PEMOLINE

In a parallel set of experiments, Abbott Laboratories researchers reported that a single administration of magnesium pemoline enhanced the acquisition and retention of a conditioned avoidance response in "slow learner" rats (Plotnikoff, 1966a) and stimulated brain RNA polymerases within nuclear aggregates in both *in vivo* and *in vitro* systems (Glasky and Simon, 1966). Plotnikoff (1966b) stated also that in rats magnesium pemoline prevented memory loss by ECS. He conjectured that this action was due in part to the acceleration of nucleic acid synthesis by magnesium pemoline and the preventing or restoring of depleted chemicals.

The retention results of Plotnikoff are reminiscent of those of Cook et al. (1963). During the extinction process the rats given yeast RNA by Cook et al. showed increased resistance to extinction. Apparently these rats adhered rigidly to a pole jumping response while the normal control animals gradually relinquished this response. Plotnikoff's methods of testing retention is basically an extinction procedure, for reinforcement is no longer present. The rats given magnesium pemoline adhered rigidly to the avoidance response, i.e., no differences in jump out time from Trial 1 to Trial 10, whereas the control group showed the usual slow decline in strength of the conditioned response. These results suggest the possibility that both yeast RNA and magnesium pemoline affect the organism in some general manner, possibly making it overly sensitive to the stimulating conditions.

Lubar et al. (1967) reported that a single injection of magnesium pemoline had a facilitative effect on performance in the Hebb-Williams maze (a test of animal intelligence). The authors were uncertain as to whether the effect was on learning and memory processes directly or was attributable to stimulation effects.

In the author's laboratory three experiments were conducted with 28 pairs of littermate white rats in a one-way active shock avoidance task (Gaito, Davison, and Mottin, 1968). The results were consistent in each experiment; rats fed magnesium pemoline prior to conditioning showed a greater mean number of avoidances in 15 trials than did control rats. Over the three experiments the means were: E, 10.8; C, 8.6. In 24 of the 28 pairs the E's had a greater number of avoidances than did C's. Such facilitative results need not be interpreted as indicating that magnesium pemoline has a primary effect on learning processes; it is possible that because magnesium pemoline is a central nervous system stimulant, this chemical may have sensitized the E rats, producing a greater reaction to the shock and indirectly led to superior conditioning. In spite of the fact that enhanced conditioning followed the administration of magnesium pemoline, increments in RNA and protein synthesis did not occur; in fact there was a tendency for lower amounts of label to be present in the cell pool, RNA, and protein fractions. This finding tends to contradict the hypothesis that magnesium pemoline stimulates RNA polymerases with a resulting increase in RNA synthesis. In these experiments the analysis of RNA synthesis was on rats involved in the shock avoidance task, whereas the Glasky and Simon analysis was with nonbehaving animals. The magnesium pemoline might have sensitized the experimental rats such that the high shock levels used might have affected these rats more than was the case with the control animals. However, there is other research which indicates that RNA synthesis is not enhanced (Morris, Aghajanian, and Bloom, 1967; Stein and Yellin, 1967). Yet Simon and Glasky (1968) later reported increased RNA synthesis in the nuclear fraction again.

In the experiments above which show that magnesium pemoline facilitates learning, the appropriate response was an active one. Thus if this chemical was merely acting as a stimulant, an active response would be favored. On the other hand, if withholding of a response (e.g., as in a passive avoidance task) is required, a stimulant would produce impaired performance. Gurowitz et al. (1967) used passive avoidance conditioning and found that a single administration of magnesium pemoline produced a deficit in avoidance behavior. These results were similar to those obtained by Gurowitz, Gross, and George (1968) with TCP. Gurowitz et al. (1967) concluded that the effect of the chemical was one of stimulation, affecting performance but not learning.

Frey and Polidora (1967) reported that a single injection of magnesium pemoline facilitated the acquisition of an active avoidance response. Further analysis indicated, however, that the facilitation was chiefly on those rats that tended to "freeze" in response to the electric shock. Beach and Kimble (1967) found that the facilitative effect could be attributable to increased activity levels and more sustained responsiveness to stimulus conditions.

The stimulant interpretation of magnesium pemoline effects is also suggested by research which indicates that this chemical improves the performance of fatigued humans for a period of at least eight hours (Gelfand et al., 1968; Herbert et al., 1968).

Other research efforts with human subjects found that a single administration of magnesium pemoline did not facilitate learning (Burns et al., 1967; Smith, 1967).

Thus the results tend to suggest that the effect of magnesium pemoline is probably one of stimulant nature, affecting performance directly and learning indirectly. Results by Plotnikoff (1967) tend to question this interpretation, however; a single dose of magnesium pemoline in rats enhanced learning capability two weeks later in an active avoidance task. Obviously, further research is required to clarify the contribution of magnesium pemoline to behavioral events. It is possible that with magnesium pemoline, as possibly with TCP, yeast RNA, and other chemicals, long-term administration is required for an effect on learning to appear.

Administration of Chemicals which Affect Protein Synthesis

Flexner, Flexner, and Stellar (1963) injected a protein synthesis inhibitor, puromycin, into the brain of mice at various time periods following the acquisition of an avoidance discrimination response (moving into a shock-free arm of a Y maze within five seconds). Puromycin interferes with protein synthesis by being incorporated into the growing polypeptide chain and causing a premature release of incomplete peptides (Harbers, Domagk, and Muller, 1968). The anatomical spread of puromycin was checked with injections of fluorescein in a number of animals. Some animals received frontal injections; others, temporal injections; and others, ventricular injections. Injections in two and three areas were employed also with some animals. All injections were of bilateral nature. The Y maze training was conducted in a single session of about 20 minutes to a criterion of 9 out of 10 correct responses. The same procedure was used to test for memory. They reported that injections after one to three days involving the hippocampus and adjacent temporal cortices caused loss of memory; these same injections after six days had no effect. Loss of memory following injections after 11 days or longer required injections in all three areas. They also found that recent reversal learning was lost with bilateral injections in the hippocampal-temporal areas but longer term initial learning was retained.

In a later study Flexner et al. (1964) found that protein synthesis in both the hippocampus and temporal cortex must be inhibited for at least eight to ten hours in excess of 80 percent for loss of recent memory of maze learning to occur. Bilateral temporal injections of puromycin achieved this degree and duration of inhibition and consistently caused loss of memory. As the concentration of the antibiotic was decreased, it became progressively less effective in its behavioral and biochemical effects so that recent memory was retained in an increasing proportion of animals as the effect on protein synthesis diminished. Subcutaneous injections of varying amounts of another protein synthesis inhibitor, chloramphenicol, did not cause 80 percent inhibition and none of the mice showed loss or impairment of memory.

Flexner et al. (1965) reported that long-term memory could be destroyed in a majority of mice by injections which inhibited protein synthesis in the hippocampus and temporal cortex by at least 80 percent for 11 hours and in a substantial part of the remaining cortex for the same time period. As the concentration of the inhibitor decreased it became progressively less effective, and the longer term memory was retained in an increasing number of animals as the effect on protein synthesis diminished. They suggested that the maintenance of memory may depend upon a continuing synthesis of protein.

Flexner and Flexner (1966) used another protein synthesis inhibitor, acetoxycycloheximide. This antibiotic alters protein synthesis by inhibiting the transfer of an amino acid from tRNA into the polypeptide chain. In spite of great inhibition of protein synthesis, no impairment of retention occurred if the heximide was given within one day or between 12 and 35 days following training. Other work in which the antibiotic was injected just before or immediately after training indicated no impairment within about five hours or after about 60 hours following training; however, there was a transient impairment of retention in the intermediate period which might have been due to the illness produced by the antibiotic (Flexner, 1967). Flexner and Flexner (1966) also reported that if acetoxycyloheximide and puromycin were injected together at one or 14 days after learning, no impairment of memory occurred.

Flexner and Flexner indicated that these results suggest that puromycin destroys memory because the rate of synthesis of mRNA is inadequate to compensate for its loss, possibly because the amount of protein (which acts as an inducer of RNA synthesis) falls below an effective level during inhibition of protein synthesis and/or possibly as a result of direct inhibition of RNA synthesis. According to the Flexners the acetoxycycloheximide prevents the disaggregation of polysomes and the degradation of mRNA caused by puromycin. As soon as the effect on protein synthesis has dissipated, the mRNA's can then function for protein synthesis.

The results of other research caused Flexner and Flexner (1967) to discard their hypothesis that puromycin destroyed memory as a consequence of destruction of mRNA's. They found that injections of saline for up to two months after treatment with puromycin one day after learning brought about a restoration, or partial restoration, of memory. However, if puromycin was injected five hours before training or four to eight minutes after training, the restoration effect by saline was less (Flexner and Flexner, 1968). Flexner (1967) and Flexner and Flexner (1967) suggested that the puromycin blocks the expression of memory, rather than destroying it, because the peptides formed are abnormal (i.e., contain puromycin) and alter in a reversible fashion the characteristics of neuronal synopses.

Barondes and Cohen (1966b), using a procedure similar to that of the Flexner group, stated that inhibition of protein synthesis by puromycin injections in the temporal lobe area of mice had no effect on the acquisition of a shock escape or avoidance response in a Y maze five hours later. Forty five minutes later, however, retention was decreased by more than 50 percent; three hours later retention was approximately 10 percent of the original level.

Barondes and Cohen (1967) repeated this study with a T maze and added cycloheximide and acetoxycycloheximide. These antibiotics produced a greater degree of protein synthesis than did puromycin. No effect occurred on acquisition five hours after training. Three hours later the puromycin mice showed impaired retention, but the mice injected with the heximides did not. The addition of cycloheximide to the puromycin solution reversed the retention deficit. Injection of cycloheximide frequently resulted in immediate death, but the mice who survived showed no further evidence of toxicity. Puromycin did not show the immediate lethal effect; however, puromycin-injected mice were lethargic for about four hours. Mice treated with the dose of acetoxycycloheximide used by Barondes and Cohen developed diarrhea within five hours of injection.

Cohen and Barondes (1967a) compared mice injected with acetoxy-cycloheximide or saline during prolonged T maze training (9 correct escape responses in 10 trials) with those exposed to brief training (3 out of 4 correct responses). All groups were able to acquire the desired response five hours after the injection and showed normal retention three hours later. For the prolonged training condition, normal retention was indicated for all groups up to seven days later. However, when brief training was used the animals injected with the antibiotic were significantly poorer in retention than the saline mice at six hours after acquisition and at various periods up to seven days later.

The previous experiments used injections bilaterally into the temporal lobes. Barondes and Cohen (1968a) used subcutaneous injections on the back during brief training (5 out of 6 correct escape responses). Mice injected with acetoxycycloheximide developed diarrhea three or four hours later and this condition persisted for a number of hours. The acetoxycycloheximide mice injected 300 minutes, 30 minutes, 5 minutes, or immediately before training, or immediately, 5 minutes, 30 minutes, or 1440 minutes after training showed a normal capacity to learn. All mice were tested for retention seven days after training. The mice injected with the antibiotic before training or up to five minutes after training had impaired retention. The time when the retention deficit first became apparent was determined by testing mice three or six hours after training. Mice injected with acetoxycycloheximide 30 minutes before learning had normal retention three hours after training but impaired retention six hours after training.

Barondes and Cohen (1968b) reported similar results with cycloheximide during brief training. They found also that the post-training presentation of an arousal agent (foot shock, amphetamine, or corticosteroid injection) given three hours after training to the heximide treated animals attenuated the memory impairment. The authors suggested that conversion of "short term" memory to "long term" memory was blocked by inhibition of protein synthesis, but that the manipulations producing arousal were successful in effecting this conversion.

In summary, the results by Barondes and Cohen indicate no effect of puromycin or the heximides on acquisition; puromycin impairs retention but the heximides produce retention deficits at certain times only when brief training is involved (3 of 4 or 5 of 6 responses correct).

The relationship between protein synthesis and learning events has also been investigated in goldfish. Agranoff and Klinger (1964) reported that puromycin injected intracranially in these fish produced impairment of memory for a shock avoidance response. Table 14 shows the data in one experiment. With control fish (Group A) the mean number of avoidance responses (i.e., moving into the dark end of a training tank in response to the onset of a light prior to the occurrence of shock) was 1.31 in the first 10 trials and 2.89 in the second 10 trials. These animals were either injected with saline or not injected, and then returned to the home tank. Seventy-two hours after the end of the twentieth trial, each fish in Group A was given 10 trials again in the training tank (retention trials);

TABLE 14   Effect of Puromycin on Shock Avoidance Responses in Goldfish

Group	N	Day 1		Day 4
		1–10	11–20	21–30
A	36	1.31	2.89	4.56
B	36	1.44	2.86	2.89

From B. W. Agranoff and P. D. Klinger, Puromycin effect on memory fixation in the goldfish. *Science,* 1964, *146,* 952–53. Copyright 1964 by the American Association for the Advancement of Science.

Group A, fish injected with saline after Trial 20 or not injected; Group B, injected with puromycin after Trial 20; results are mean number of avoidances in 10 trials.

the mean number of avoidances was 4.56. Notice that the level of learning increases slowly over the three sets of trials. Notice also that level of learning is quite low even during the last block of trials (i.e., a mean of 4.56 avoidances in 10 trials) as compared to the learning criterion imposed by the Flexner and Barondes groups (9 of 10; 5 of 6; or 3 of 4 responses correct). The experimental fish (Group B) were injected with puromycin after the twentieth trial. Seventy-two hours later these fish had a mean of 2.89 in the 10 retention trials.

There were no significant differences between Groups A and B in the first 20 trials. However, the two groups did differ in the retention trials. Agranoff and Klinger interpreted these differences as indicating an impairment of memory because of the puromycin. If impairment of memory had occurred, one would expect that Group B fish would have shown the same number of responses during the retention trials as they did during the first 10 trials. This is not the case, however; their performance is actually the same as during the second 10 trials; thus it appears that a lack of improvement may be involved rather than a memory deficit. Unfortunately, in these 10 trials learning and memory (or acquisition and retention) are confounded, because of the low level of learning attained by the end of the twentieth trial.

Agranoff and Klinger also reported that if goldfish were trained over a period of days until they were responding correctly 80 percent of the time with no further improvement occurring, no differences appeared between control and puromycin fish in the retention trials.

Further results were obtained in a number of other studies (e.g., Davis and Agranoff, 1966; Agranoff, Davis, and Brink, 1966; Agranoff and Davis, 1968). They maintained that injections of puromycin just before the first trial or within one hour after the twentieth trial adversely affected performance during the retention trials. In these studies they used a predicted score based on a regression analysis developed on a large control group. The retention score for trials 21–30 was the observed score minus the predicted score. This procedure can precipitate an odd result; when performance over the three blocks of trials is essentially the same, the retention score will indicate a decrement in trials 21–30. For example, Agranoff, Davis, and Brink (1966) reported that in the three blocks of trials the observed mean scores for a puromycin group were 3.0, 3.8, and 3.3; these values would appear to be essentially the same, with no improvement occurring over the three blocks of trials. However, the predicted score for the reten-tion trials is 5.5; thus the retention score is −2.2 (i.e., 3.3 − 5.5). In this case the prediction score may be obscuring an important aspect, i.e., that the fish do not improve, not that they show a deficit in memory.

Agranoff, Davis, and Brink (1966) reported that injections of large doses of puromycin resulted in an observed score on trials 21–30 which was the same as that obtained on trials 1–10; they also found that acetoxy-cycloheximide produced a memory deficit (i.e., a negative retention score). This result is contrary to the results of the Flexner and Barondes groups who found no effect of this antibiotic when using a learning criterion of 9 correct in 10 trials (prolonged training), but is consistent with the results of Cohen and Barondes (1967a) with a criterion of 3 of 4 or 5 of 6 responses correct (brief training).

Agranoff and Davis (1968) stated that maintaining the fish in the training tank following Trial 20 (rather than returning the fish immediately to the home tank) had an effect on performance during the retention trials. Fish remaining for 1.5, 3.5, 24, and 48 hours in the training tank and then given an injection of puromycin showed retention deficits on Day 4 which were not significantly different from that obtained in fish who were injected immediately after Trial 20. Uninjected controls showed no deficit in reten-tion scores following 1.5 hours in the training tank after Trial 20; if main-tained 3.5 hours in the training tank before removal to the home tank a slight retention deficit resulted. Groups maintained for 24 or 48 hours in the training tank showed retention score deficits which were equivalent to those of the puromycin groups. The authors maintain that these results indicate that fixation of memory is suppressed by stimuli in the training tank other than the conditioned stimulus (light) and the unconditioned stimulus (shock). It is possible that the period in the training tank follow-ing Trial 20 is allowing an extinction or habituation effect to occur.

An interesting study by Oshima, Gorbman, and Shimada (1969) utilized EEG recordings of salmon olfactory nuclei to detect recognition of "homing" waters, a very specific form of animal memory. Appropriate quantities of either puromycin, cycloheximide, or actinomycin D were given. Animals were tested four to seven hours later for recognition of home water. The salmon were incapable of recognizing home water under any of the three drugs. One implication is that continuing protein and RNA synthesis is necessary for memory "readout" to occur, as memory was again detectable in these fish 24 hours after drug administration. Alternatively, these results may indicate that memory "readout" is adversely affected by these drugs but becomes available when the drugs are no longer functional. The Flexner and Flexner results (1967) with saline injections and the Barondes and Cohen results (1968b) with arousal agents are consistent with this interpretation.

Of the research using protein synthesis inhibitors, only that using puromycin shows a consistent impairment of memory over various conditions. However, this research does not indicate whether the effect of this antibiotic on memory is of primary, secondary, or parallel effects nature. That the effect is not a primary one is suggested by a number of research results. Cohen, Ervin, and Barondes (1966) checked the electrical patterns in the hippocampal region during the operation of puromycin and cycloheximide. With the former there was an attenuation of electrical patterns. The electrical activity of mice injected with cycloheximide was indistinguishable from activity of saline injected mice. Thus they reasoned that the amnesic effect of puromycin involved other mechanisms than its effect on protein synthesis. Research results which are consistent with this interpretation have been provided by Avis and Carlton (1968). A marked decrease in the amplitude of hippocampal activity was produced by an injection of potassium chloride 24 hours after learning; attending this diminution of electrical activity was a deficit in retention. Cohen and Barondes (1967b) also found that puromycin increased the susceptibility of mice to seizures, whereas acetoxycycloheximide or both antibiotics together did not.

Gambetti, Gonatas, and Flexner (1968) found that puromycin produced swelling of neural mitochondria whereas acetoxycycloheximide did not. Puromycin with acetoxycycloheximide resulted in minimal swelling. The authors suggested that the deleterious effect of puromycin on memory was due to action on nerve cell membranes.

Other general cellular impairments due to puromycin have been reported by Appleman and Kemp (1966) and by Jones and Banks (1969). The former authors reported drastic decrements in energy metabolism which was independent of its effect on protein synthesis; the latter group found decrements in tissue respiration.

The major problem with the indirect approach of this and the previous chapter is the relative uncertainty concerning the effects of the administered substances on the total chemical environment of the organism and the relationship to the behavioral events (primary, secondary, or parallel effects). While many researchers assume specificity of drug action sites, such specificity probably is rare. It is often assumed, for example, that actinomycin D is a specific inhibitor of RNA synthesis, while puromycin and cycloheximide are potent protein synthesis inhibitors. Yet these three drugs have the same net effect, at least in part, in the prevention of ribosome maturation; this in turn leads to a deficit in protein synthesis, and little or no mRNA reaching the cytoplasm. Thus inhibitors of RNA and protein synthesis may have a number of sites of action but produce similar effects through separate modes of action.

On the other hand a chemical can produce diverse effects; for example, as indicated above, puromycin causes an attenuation of electrical activity in hippocampal cells, increases the susceptibility of mice to seizures, produces swelling of neural mitochondria, and reduces energy metabolism and tissue respiration, in addition to inhibiting protein synthesis. Such diverse effects probably are characteristic of other chemicals which affect RNA and/or protein synthesis or degradation. Both RNA and protein are so intimately concerned in general cell functions as to render any conclusion about specific effects very tenuous. To attempt to separate a single specific effect from the overall general effect is like "looking for a needle in a haystack."

Because of this problem the results with the indirect approach must be viewed with caution. As stated previously, the indirect approach is best used as a supplement to the direct approach.

Chapter 10

RESEARCH RESULTS: CRITIQUE

The research results of the previous chapters on the DNA complex were concerned with a narrow segment of the overall problem of behavior and brain function. However, an understanding of these intracellular phenomena may help to suggest other events which are transpiring.

The results suggest that DNA derepression or activation may occur in some adaptive behavioral events and that RNA synthesis and subsequent protein synthesis may be directly or indirectly involved in these behaviors. Whether histones are the repressors of DNA is not certain. The research discussed in Chapter 4, however, does suggest this as a plausible possibility.

Whether RNA and protein serve as important molecules in learning phenomena is not certain. There are two assumptions by those who assign an important role (i.e., primary effects) to these molecules during learning.

Assumption A: During learning, qualitatively different RNA and protein molecules are synthesized which are unique in the history of the neural tissue involved, or are unique to the learning behavior.

Assumption B: These molecules have a primary effect in mediating the production of the necessary behavioral responses to the provoking stimuli (i.e., the presence of these molecules does not indicate merely a secondary or parallel effect). (See the discussion in Chapter 6 on primary, secondary, and parallel effects).

Most of the research techniques have been too gross to detect possible qualitative changes. The "base changes" in RNA reported by Hydén (Chapter 6) may not indicate new RNA species being synthesized during learning but may be a statistical artifact because of the pooling of many RNA's; the same species may be synthesized but in different proportions than in the control situations. The research approach which is most directly related to Assumption A is that conducted in the author's laboratory using DNA-RNA hybridization procedures. Further research with these techniques is required, however, before one can say *conclusively* that *unique* RNA species occur in learning. These procedures when carried further may assess also the question of the uniqueness of protein for behavior.

Although some researchers offer both Assumptions A and B, it is possible that Assumption A could be false (i.e., no unique RNA present during learning) but Assumption B, true. That is, increases in amounts of RNA may be enough to mediate the behavioral responses during learning.

There are a number of points which one should keep in mind when evaluating research efforts relative to Assumption B. They are:

1. It is not possible to "read" from the brain of an animal during and following the acquisition stage to determine what changes have occurred. Therefore, one must resort to performance measures to assess learning phenomena. This requirement means that performance parameters are confounded with learning aspects (acquisition and/or retention), and it is difficult to consider the latter even though one may attempt experimentally to estimate the performance contribution with certain controls. This confounding causes difficulty in the interpretation of experimental results. For example, RNA and protein synthesis inhibitors have a gross effect on cellular functions such that one would expect that the performance of an animal would be affected adversely; indirectly learning capabilities might be modified also.

A related point is that in assessing the presence of memory by testing retention, there is a confounding of the memory and retrieval systems (assuming that they are separate). Memory may be present but the retrieval system may be disturbed, leading to an erroneous conclusion that memory is absent. This confounding of memory and retrieval systems might explain the results of Flexner and Flexner (1966). They found that when mice were injected intracerebrally with acetoxycycloheximide there was an initial period during which memory was retained, an intermediate period in which expression of memory was lost, and a final period during which memory returned. The results of Flexner and Flexner (1967) with saline injections and those of Barondes and Cohen (1968b) with arousal agents may be showing effects on the retrieval system.

2. The results with the RNA and protein synthesis inhibitors are concerned with synthesis events. It is possible that an important event during learning behaviors may be a reduction in degradation of RNA and/or protein which would allow these species to function for longer time periods. Research with plants has indicated that hormones increase the resistance of some RNA species to degradation by RNase (Bendana and Galston, 1965). The neural events during learning may operate in a similar fashion. Investigators will need to determine the amount of RNA, protein, and RNA and protein constituents as well as the synthesis patterns which are occurring for a more complete picture of functional activity.

Consistent with this point is the possibility that redistribution within the cell of existing RNA and/or protein species may be important events for learning phenomena. Cohen and Jacklet (1965) emphasized that changes in the distribution and concentration of RNA in the cytoplasm of vertebrate central neurons were related to a variety of different functional states in these cells.

Unfortunately, as with Assumption A, the research results are inconclusive relative to Assumption B. The Hydén and Glassman results, and others, suggest that Assumption B may be correct; however, their procedures do not eliminate the possibility that the RNA changes detected exemplify the secondary or parallel effects possibilities rather than the primary effects one. The experiments in which RNA or homogenates are extracted from trained animals and then injected into naïve recipients are important ones, and are the closest to showing that primary effects are occurring; however, many contradictory results abound. Likewise, the research results involving the administration of various chemicals such as magnesium pemoline, actinomycin D, TCP, etc. are not clear. Such chemicals probably influence many chemical reactions and have diverse effects, thus making difficult a differentiation between primary, secondary, and parallel effects possibilities. The research using protein synthesis inhibitors appeared at first to be less controversial; however, here there is also the problem that these antibiotics affect other processes than protein synthesis and may be showing parallel effects. Thus no research effort to date has been successful in differentiating the RNA and protein changes, noted during many behaviors (Gaito, 1966), which are of secondary or parallel nature from those which may be primary effects. Unfortunately, the methods which have been used are too gross to allow such a differentiation.

Pevzner (1966) thought that RNA changes during many behaviors represent secondary effects. He cited work by Brodsky concerning light adaptation. RNA synthesis was observed during this event; however, the RNA synthesis followed, rather than preceded, the neural functions in light adaptation. Brodsky concluded that the RNA changes probably reflected a secondary process and that the physiological activity of the neurons was based directly on other neurochemical reactions. Pevzner thought that the question "Is the role of RNA in behavior a primary or secondary one?" was a basic one in evaluating research results.

Thus, based on the technology and research results available at this time, it is not possible to say that Assumptions A and B are either true or false. However, because at least seconds or minutes are required for protein synthesis to occur, it appears likely that the acquisition phase of learning does not require the presence of RNA and/or protein because learning sometimes occurs almost instantaneously. Therefore, if RNA and protein are important molecules in mediating behavior, their influence probably would be on memory (retention).

It is obvious that the research efforts concerned with the DNA Complex and its possible role in behavior are exciting and of profound importance; at this time it is difficult to provide definite conclusions. This field is just developing and is in a state of fluidity. However, important results should be provided in the near future.

Chapter 11

MOLECULAR MODELS OF LEARNING

Having discussed the DNA Complex, and research results relative to this complex, let us consider now some models or theories which suggest how this complex, or portions of this complex, function during behavior. The behavioral event which has been emphasized is learning. As with the research results, the models have emphasized the intracellular events and thus are focusing on a narrow segment of the overall problem.

Earliest Modern Approach

Up to approximately twenty years ago protein molecules were considered to be the genetic substances. Protein was also the favorite candidate for being involved in the memory trace (Katz and Halstead, 1950; Halstead, 1951; Gerard, 1953). Although a number of individuals suggested the possible involvement of proteins in memory, the first systematic set of hypotheses was by Halstead. Katz and Halstead (1950) and Halstead (1951) suggested that proteins in the cell nucleus (nucleoproteins) were the substances which had the ability to act as templates on which replica molecules were formed. At first the neurons of the brain were supposed to contain random configurations of protein. Stimulation of neural tissue by nerve impulses caused the randomly oriented molecules to assume a specific configuration. Nucleoproteins were involved in these reorientations and became templates. They believed that these templates were like those representing native endowment but differed from them in arising from external stimulation. They stated that the ordering of the protein templates could take place in various components of the cell and its processes, including the synapse. However, the reorganized protein replicas ultimately resided in the neural membranes where they participated as "traces." The linking together of numerous neural units for "memory" occurred only between those units with similar nucleoprotein configurations.

These hypotheses stimulated some interest among scientists, but the possibilities of the suggested model were not recognized at that time. Apparently the intellectual climate was not suitable for these ideas. However, in the intervening years with the important biological discoveries relative to the nucleic acids, the attention of psychologists and other behavioral scientists has been directed again to the molecular level.

If one substitutes the current molecular favorites, DNA or RNA, for the protein in the Halstead hypothesis, he sees that some of the current molecular hypotheses are similar, if not identical, to those of Halstead. Thus it is evident that Halstead cogently anticipated the current thought.

Instructive Models—Qualitative RNA Changes

In discussing the various molecular models of complex behavior, theorists have tended to borrow several terms (instructive, selective) utilized in the interpretation of antigen-antibody reactions. The instructive approach to this event maintains that antigens in some manner shape each antibody whereas the selective idea is that certain cells have the potential for synthesizing any antibody and that the antigen merely releases that potential.

Holger Hydén has been the pioneer in proposing and evaluating the hypothesis that RNA is of basic importance in learning. At first, he advocated an instructive model which specified that stimulation modified RNA molecules (1959, 1961), but his recent work suggests that he has shifted to an emphasis on a selective model. In his early theory, he hypothesized that memory is a result of a change in the sequence of bases in the RNA molecule. The first series of impulses generated in sensory cells or in motor neurons pre- or postnatally change the stability of one or more of the four bases of the RNA molecule at a certain site along the molecule. Precisely which changes will occur depend on the frequency of the first series of impulses generated in the nerve cell. This effects a change of one against another base from the cell pool. The new base at this space is now stable under the influence of the frequency pattern. Since the sequence of the bases in the template RNA is now changed, new protein formed through the mediation of this RNA will also be specified.

Later stimulation causes a rapid dissociation of the specified protein and a subsequent combination of the dissociated products with a complementary molecule. Through this rapid combination of the dissociated protein with its complementary molecule, an activation of the transmitter substance occurs, and the postsynaptic structure is excited.

These modifications would occur in numerous cells in the brain. The ultimate "memory trace" would be maintained by the interrelationship of brain units via these changes. Presumably the neural aspects of the Hydén model might be somewhat similar to those of Hebb (1949), with some modification.

Thus, according to Hydén, a nerve cell responds differentially depending on whether the pattern of impulses it receives is novel or familiar, as well as on the specific pattern itself. No protein will have the correct configuration if the incoming pattern of impulses is new; therefore, no dissociation of protein can occur. The electrical pattern must first shape a new RNA molecule, which in turn shapes a protein molecule that can dissociate. If, on the other hand, the incoming impulse is familiar, protein molecules will already be present that are capable of dissociating rapidly.

The author's interest in Molecular Psychobiology began in late 1959 upon reading the excellent articles by Crick (1954, 1957) in the *Scientific American*. He saw immediately the possibility of having heredity and environment operate in one specific molecular structure, DNA. A theoretical paper discussing these possibilities was prepared, which was received and accepted by *Psychological Review* in early 1960. In this article (Gaito, 1961) he advocated an instructive model, but one concerned with DNA base changes. He suggested that changes might occur at the attachment of the two strands of DNA, with A at one locus changing to G and the associated pyrimidine changing from T to C. The two purines, A and G, are quite similar in structure; the only differences are in the side branchings; the same is true of the pyrimidines, T and C (see Figure 4, Chapter 2). Thus a small change in one purine, or pyrimidine, could lead to the other one. Deletion, addition, and rearrangement of bases were considered as other means of changing the code. These changes would provide a basis for modification of the genetic potential in nerve cells by means of external stimulation during learning, which would be a somatic mutation not transmitted to the offspring. Other possible biochemical mechanisms also were suggested which could be involved in memory functions, i.e., changes in RNA or amino acid sequences.

Dingman and Sporn (1961) hypothesized that RNA changes were the basis for memory. They suggested, however, that the linear sequence of bases (primary structure) was only one possible means of coding experiential events; changes in the helical structure (secondary structure) and overall configuration (tertiary structure) also could be the basis for memory.

Instructional models have been suggested also by Cameron (1963), Gerard (1963), and by McConnell (1964).

Selective Models—Release of Specific RNA

Smith (1962) stated that the enzyme induction model might be suitable for learning events. He suggested that the inducer substance may be acetylcholine (ACh) which is released at the synapse during stimulation. The proteins then induced via RNA synthesis would be the enzyme required for the synthesis of ACh, choline acetylase (ChA), and acetylcholinesterase (AChE), the enzyme which degrades ACh. These events were presumed to increase the amounts of ACh functioning at the synapse within a given time period and increase the probability that stimulation in neural units would activate other neural units, thus leading to greater potential for adaptive behavior. Briggs and Kitto (1962) and Goldberg (1964) expressed somewhat similar views.

A recent approach to learning by Hydén (Hydén and Lange, 1965, 1966) is that DNA sites in glia and in neurons are stimulated such that unique RNA species are synthesized. During the early part of the learning process, the RNA that is synthesized is high in A and U, thus being DNA-like in composition. This RNA is formed during the establishment of functional synapses for the new behavior, and this stage represents short-term memory. During the later portion of the learning process an RNA rich in G and C (similar to ribosomal RNA) is formed. This stage is supposed to constitute the fixation of long-term memory with a high synthesis of transmitters at the synapse. Although Hydén favored a selective model for most learning, he suggested that insightful learning and problem solving would involve an instructive mechanism (1967). The exact mechanisms, however, were not specified.

Other selectional approaches were the DNA derepression model of Bonner (1966) and the DNA activation model of Gaito (1966). Bonner suggested that in neurons there is a gene, or a few particular genes, which are repressed, but which are derepressable by certain substances which are released as a result of electrical stimulation of specific portions of nerve cells, viz., dendrites. Once derepressed the gene makes more RNA and ultimately more enzyme. The enzyme then is involved in the chemical reaction which makes more of the substance which derepresses specific repressed DNA sites. These genes, once derepressed, remain derepressed permanently. Such derepression would account for the increased rate of RNA and protein synthesis which is reported in learning experiments.

The DNA activation model hypothesized that during learning events, stimulation of specific nerve cells in a particular portion of the brain caused a modification in the DNA complex such that DNA is activated to synthesize RNA. Four types of RNA were assumed to be synthesized: messenger(s) for synaptic and nonsynaptic protein, messenger(s) for ribosomal protein, messenger(s) for enzymes to synthesize lipids, and ribosomal RNA. The protein proceeds to synaptic junctions to increase the surface area, to make for the development of connections with other nerve cells, and to make postsynaptic neurons more susceptible to excitation. The lipids are incorporated along with proteins into membranes at the synapse and elsewhere. The ribosomal RNA and protein aggregate to make new ribosomes. The synaptic changes link neural circuits together to make more probable the inclusion of certain cells in the electrochemical stimulation which results from the physical energies impinging upon the receptors of the organism when external stimulation occurs. This stimulation allows the DNA complex in a greater number of nerve cells to be potentially available for modification so that further protein synthesis can occur. Since ribosomes function as vehicles for protein synthesis, the increased number of ribosomes would provide for a more rapid rate of protein synthesis during later stimulation.

Flexner (1967) offered a selectional model involving a self-inducing or self-sustaining system. The learning experience triggers the transcription of mRNA, which gives rise to protein essential for the expression of memory. This protein, or products derived from it, serve as inducers of its related mRNA.

Critique

The Halstead approach would be classified as an instructive model because the emphasis is on stimulation causing the randomly oriented molecules to assume a specific configuration. These molecules with their modified configuration then would be templates for the formation of other nucleoproteins with comparable configuration. This model represented an integration of protein chemistry with learning research results in an ingenious fashion.

The instructive model with the idea of nucleic acid base changes as suggested by Hydén, Dingman and Sporn, and Gaito seemed to be inconsistent with basic conceptions in molecular biology. The main difficulty was in the mechanism required in effecting base modifications. Changes in bases can occur but require mutagenic agents such as nitrous acid, ultraviolet light, etc. Invoking a mechanism in learning similar to mutagenic agents would be illogical. Thus these individuals shifted later to selective hypotheses.

One cannot exclude completely the instructive model as a possibility because some of the experimental results may be consistent with it; however, there are multiple interpretations for each such result. Furthermore, the fact that Hydén and others who favored this viewpoint at one time have abandoned it argues against its plausibility. The selective model is more attractive because it is most consistent with molecular biological research results.

Recently a model was suggested which incorporated both instructive and selective mechanisms. Griffith and Mahler (1969) speculated that base changes (via methylation or demethylation) in specific DNA sites could determine the number of times that an induced mRNA would be used in protein synthesis. Each mRNA was assumed to have a number of RNA tickets adjacent to the coding portion. Only the mRNA's with base changes in one or more RNA tickets (via the base changes in complementary DNA sites) would result in protein synthesis. A protein or polypeptide would be synthesized for each modified ticket. This model is reminiscent of the earlier Gaito (1961) model, but it does incorporate the selective model which is widely accepted for molecular biological phenomena. At this time, however, the model must be considered highly speculative, as is the case with any of the models. Research within the next few years should provide a better basis for evaluation of these models and for determining if RNA and protein function as primary effects in learning, an assumption which appears to be common to all of the above models.

Chapter 12

DNA COMPLEX THERAPY

Description

The DNA Complex and its role in behavior is so basic that the applicability of information regarding this aspect would appear to be relatively unlimited. In Chapter 1 the reader was introduced to some of the exciting developments occurring in molecular biology involving this complex. Obviously, these aspects could be extended to the adaptive behavior of concern to Molecular Psychobiology.

There seem to be three broad ways in which adaptive behavior might be influenced. The most drastic means would be by modifying the DNA of an organism. This aspect has been discussed in Chapter 1. Scientists have been able to make active DNA and RNA in the test tube by modifying viral nucleic acids. It seems possible that eventually molecular biologists will have the capability of synthesizing DNA (or modifying the DNA within an individual) which would provide the potential for specific characteristics such as increased longevity, better physique, higher intelligence, increased frustration tolerance, and other aspects. This possibility requires the knowledge of the specific DNA codes for these characteristics. Such information will not be available for a number of years. A complicating aspect is that social, religious, and moral problems arise when one thinks of modifying the genetic structure of human organisms. A major problem would be opposition to the modification of life by social, religious, or other moral organizations. One society or organization might consider certain characteristics as "desirable" ones in contradiction to those of others. However, this possibility is an exciting one from a scientific viewpoint and molecular biologists should make rapid progress toward such a goal.

In contrast to the first means of influencing behavior by modifying the genetic potential of human organisms, the two other possibilities are not contaminated with the same social, moral, and religious problems; nor are they as basic and drastic. These possibilities involve actualizing the genetic potential of the organism so as to facilitate improved performance in various behavioral events.

If scientists are able to determine the mechanisms whereby specific DNA sites are turned on to produce specific RNA's, then chemicals which affect these mechanisms could be administered to the organism. Hormones appear to function in this fashion to provide for the "physiological well-being" of the organism. With further research the regulation of DNA sites which are concerned with intellectual, emotional, and performance aspects may be elucidated. A fascinating possibility which seems feasible would be the turning on of certain genes early in life which usually function at a later age; thus the developmental pattern would be accelerated in certain respects such that the organism would be more receptive to external stimulation.

The third possibility is that of determining the gene products (RNA, protein, or other chemicals) which are related to increased capability (emotional, intellectual, performance) by the individual and administering these products in appropriate form.

The last two possibilities are within the realm of behavioral scientists, and rapid developments should occur so as to improve the overall behavior of human organisms. It is important to emphasize, however, that these improvements in behavior would represent a *facilitation of behavior rather than a transfer of "information."* Some researchers (Chapter 8) talk of a transfer of "memory" or "information" from trained animals to naïve ones. It seems unlikely that such transfer does occur. What is more likely, however, is that specific chemicals may release potential capacities of the organism (as illustrated, e.g., by chromosomal puffing in insects with ecdysone) or modify the inter or intracellular environment such that more efficient conditions prevail.

The last possibility would seem to be the easiest to effect and to control within the near future. The reported improvements in performance via administration of RNA, TCP, and magnesium pemoline discussed in Chapters 8 and 9 may illustrate this possibility; however, performance improvements do not occur consistently nor are the mechanisms involved clearly understood.

Other research results which may illustrate this possibility are those by McGaugh. McGaugh and Petrinovich (1965) and McGaugh (1966) reported that a number of chemical agents have increased the rate of acquisition and stabilizing of a response in a learning task. These agents include strychnine sulfate, picrotoxin, and 5, 7-diphenyl-1, 3-diazaadamontan-6-ol (1757 I.S.). They hypothesized that these central nervous system stimulants improve learning by hastening the "consolidation of memory" during the early phases of learning. Presumably, if "consolidation" is hastened, one would expect that RNA and protein synthesis might be involved in this event.

Some studies (e.g., Heinbecker and Bartley, 1939) found that strychnine sulfate injections increased central nervous system excitability and lengthened the interval during which successive stimulations could produce persistent cortical activity. A preliminary study in the author's laboratory was concerned with the chemistry of the rat brain at various periods following IP injections of strychnine sulfate. The results indicated that at 15 minutes after injection the amount of amino acids in the cell pool was lowest and that the ratio of protein to amino acids was highest. This time period was within the period which McGaugh and Petrinovich found as being facilitative for learning. These results suggest that strychnine sulfate may reduce protein degradation events during the first 15 minutes after injection such that the relative amount of protein per amino acid is greatest at 15 minutes. This may mean that protein is more resistant to degradation as a result of administration of strychnine sulfate and that protein molecules can function longer during fixation of the memory trace.

Application to Mental Health

Mental health problems of major or benign nature are of concern to everyone. In the past, agents such as electric shock therapy, metrazol therapy, and surgical therapy have been used with severe problems such as occur in the psychoses. Tranquilizer therapy has been of use with milder cases. Psychotherapy is a treatment for both severe and mild problems. Unfortunately with all of these therapies some individuals are helped whereas others are not. Thus a chemotherapy based on knowledge of the DNA Complex might provide a greater probability of success.

The reported use of RNA therapy by some researchers (Cameron and Solyom, 1961) in alleviating emotional problems in senile and arteriosclerotic patients indicates some benefit that may be derived from using DNA products. Similar results eventually might be obtained using DNA products with younger individuals who do not have the severe problems of the aged. Favorable results may be obtained with relatively normal persons in meeting everyday problems. Tranquilizers serve this function at present; however, with further knowledge of the function of the DNA Complex in behavior, more suitable chemical agents might be efficacious.

Application to Intellectual Aspects

The process of learning within and outside formal educational systems is of utmost importance to every individual. Each person appears to have a certain intellectual potential based on his genetic structure. The external and chemical environment must allow the individual to develop his potential. Unfortunately, it is probably true that most individuals do not approach or reach their potential.

In determining the best means for developing the intellectual potential, one must control the emotions and distractions, arouse attitudes and interests, and present the material to be learned in suitable fashion. Many psychology or education textbooks present learning principles to aid in maximizing the learning process. These principles might be supplemented by a chemical approach. If scientists are able to determine the function of the DNA Complex during the learning process, the administration of appropriate chemicals should facilitate the attainment (or near attainment) of the intellectual potential for most individuals. The studies reported in Chapters 8 and 9 and those by the McGaugh group discussed in this chapter illustrate a chemical approach to the control of the learning process. Some of these appear to be successful.

A more pressing problem is with the intellectually handicapped. If scientists could determine the RNA's or proteins, or other chemicals, which some intellectual retardates may lack, one could administer these to the retarded and expect higher intellectual and other performance. Or other chemicals which are determined by these RNA's or proteins could be administered. If one could solve partially the chemical inadequacy problem of some intellectual retardates, the discovery would greatly reduce the suffering of this small segment of the human population.

The RNA therapy of Cameron and Solyom (1961) was concerned originally with alleviating intellectual and perceptual handicaps in elderly individuals.

Application to Industrial and Related Performance Problems

Each individual spends many hours of his life engaged in work performance. Thus the possibility of improving such performance would be of great value to man. Human Factors or Human Engineering has been concerned with maximizing harmony between man and his work environment so that performance of his duties will be at a high level. The chemical approach aimed at improving performance would be a valuable supplement to human factor principles for maximizing work performance.

Knowledge of the operation of the DNA Complex during motor activity, intellectual tasks, and other behaviors required in the work environment might provide a basis for substantial improvement in work performance, a savings in motor activity or other aspects by the individual, and a savings in money by management. The discussion above concerning the facilitation of the learning process would be pertinent here. For many jobs, the individual requires extensive training before he reaches his peak of productivity. Thus administration of appropriate chemicals might shorten the training process.

Such chemical therapy might be suitable in military tasks also. Appropriate chemicals might be administered to pilots, astronauts, sailors, infantrymen, etc. prior to a mission so as to facilitate greater performance during the task of concern. These developments would result in positive chemical agents which improve the capability of military forces (as contrasted with negative chemical agents which reduce the enemy's capabilities).

Application to Aging

When one reaches the middle and advanced ages, physical and mental capabilities tend to deteriorate. Probably modifications at the DNA Complex level cause, or at least accompany, these changes. Harbers, Domagk, and Muller (1968) reported that aging appears to have some effect at this level; the template activity of DNA for RNA synthesis seems to decrease with increasing age. Furthermore, it is generally more difficult to extract DNA in high yields from organs of older mammals than from those of younger animals. These observations might suggest that metabolic changes due to aging may be attributed in part to an "exhausting or wearing out" of the DNA template.

Gordon (in press) was concerned with the possibility that the aged tend to show more errors in the pathway of biochemical information flow, specifically in the machinery responsible for the translation of genetic information into proteins. He used an ultraviolet absorbance ratio as a measure of hydrogen bonding in polysomes with aged (24 months) and young (3 months) rats. In the aged rats the smaller polysomes of the brain or liver showed greater hydrogen bonding, and larger polysomes, less hydrogen bonding, than occurred in the polysomes of the younger animals. With an increase in temperature to 37°C, the RNA base hydrogen bonding in aged brain polysomes showed a greater change than did those from younger brains. He concluded that the bond types which contribute to the internal structure of polysomes was significantly different in an aged brain than in young brains, leading to a marked deficiency in the capacity to synthesize protein.

Gordon reported that administration of diphenylhydantoin (DPH) to aged animals appeared to affect the brain polysomes, moving them towards the young normal state; concurrent with this result was a partial restoration of the deteriorated capability to learn and to remember which was present in the aged animals. He indicated also that preliminary work suggested a similar improvement in behavior in humans with DPH.

Gordon used a second drug, NP-113,* with aged animals. It tended to

* Developed by Newport Pharmaceuticals, Inc., Newport Beach, California.

restore young normal characteristics to the age-altered conformation of RNA in brain and liver polysomes, enhanced the synthesis of RNA and protein, and improved the learning and memory functions of these animals.

Hydén (1961) dissected motor nerve cells in the spinal cord from humans killed in accidents and found that the amount of RNA per cell increased to about 50 years and declined thereafter. Hydén (1967) reported similar results in brain cells of rats. Related to these results is a report by Kral and Sved (1963) concerning RNase activity in the blood. They investigated the activity of this enzyme for individuals from about 20 to 100 years of age. The activity at 60 years was about 25 percent greater than at age 20 and at 100 years, about 50 percent greater. If the increase in RNase activity in blood is representative of enzymatic activity in the brain, and if spinal cord RNA changes exemplify brain RNA events, then one might suggest that the amounts of RNA decrease in old age because of RNase activity. The decrement, however, might be due to the reduced efficacy of the DNA template. Or both events might be present.

If the DNA template is less effective in aged individuals, one might consider providing "young" DNA for the cells by oral administration or by injections. Cameron and Solyom (1961) found that administration of DNA did not improve the memory or general well-being of aged humans. However, the DNA was not from human beings; thus it is possible that DNA extracted from the brain tissue of young individuals might stimulate RNA synthesis in aged persons and lead to increased capabilities.

Yeast RNA administration to aged individuals (Cameron and Solyom, 1961) did produce improvement in memory and in general processes, but RNA therapy had to be continuous or improvements were lost. Other therapeutic attempts have not provided improvements. It is possible that certain RNA species from humans might be more efficacious in providing for enhanced performance.

Alternatively the DNA complex in aged individuals might be modified by appropriate chemicals such that the DNA template could proceed more effectively for RNA and protein synthesis.

The examples discussed in the sections above illustrate a few of the possibilities with DNA Complex Therapy. Obviously much research will be required before such therapy can be successful. However, rapid developments should occur in the next decade. The future appears bright for such endeavors.

GLOSSARY

Biological—Chemical

Acetylation The process of incorporation of acetate into certain parts of a molecule.

Amino acids The units of which proteins are constructed. Two amino acids join to form a peptide. A number of peptides is a polypeptide. The difference between a polypeptide and a protein is one of degree; the latter contains more amino acids than the former.

Angstrom (Å) A minute unit of length equal to one ten-thousandth of a micron or one hundred-millionth of a centimeter.

Antibiotic An antibacterial substance produced by a living organism. For example, actinomycin-D and puromycin are antibiotics which inhibit RNA and protein synthesis, respectively.

Antibody A special protein synthesized by an organism when an antigen enters an organism for the first time. The antibody reacts against the antigen to prevent the latter from exerting a harmful effect.

Antigen A substance foreign to an organism which elicits the synthesis of a special protein called an antibody when the substance enters the organism.

Bacterium A single cell microorganism which is the smallest of living cells.

Bacteriophage A virus which infests a bacterium; also called a phage.

Brain stem The upward continuation of the spinal cord in the brain area on which the cerebral hemispheres are attached. The brain stem consists of four portions: medulla, pons, mesencephalon (midbrain), and diencephalon.

Cellular differentiation The process whereby a generalized cell becomes a specialized cell, e.g., liver cell, kidney cell, neuron. DNA regulation apparently occurs to turn on certain genes and turn off other genes in each specific cell.

98

Cerebral cortex The outer layer of the cerebrum containing gray matter (nerve cells). This cortex is the most recently evolved of the various brain structures, and in man is well developed such that it consists of a series of invaginations or wrinkles.

Cytoplasm The portion of the cell outside the nucleus.

Degradation The splitting of a molecule into its constituents, e.g., RNA into nucleotides.

Denaturation To change a substance from its natural state. For example, double stranded DNA is denatured into single strands if heated at about 95°C for 10 minutes.

Enzyme A special class of proteins whose presence affects the rate of a chemical reaction.

Escherichia coli (E. coli) A bacterium which is used in much of molecular biology research. It infests the gastro-intestinal tract of humans.

Fungus A group of plants such as molds, mildews, mushrooms, etc.

Glia Nonneural supporting cells in the nervous system. Some individuals such as Hydén have assumed that these cells play an important role in memory events.

Hemoglobin A substance in the red blood cells which combines with oxygen.

Incorporation Becoming part of a molecule. During the synthesis of a molecule, e.g., RNA, radioactive precursors such as uridine-5-H^3 are taken into the newly formed molecule.

Intraperitoneal (IP) injection The injection of a substance into the peritoneum, the membrane that lines the cavity containing the digestive organs and other viscera, so that it may spread to organs throughout the body, especially the brain.

In vitro Occurring outside the organism, e.g., in a test tube; contrasted with *in vivo,* which indicates within the organism.

In vivo Occurring within the organism; contrasted to *in vitro,* which indicates outside the organism.

Liquid scintillation spectrometry The use of a spectrometer to detect disintegrations of radioactive materials in a specific liquid medium.

Macromolecules Large molecules whose molecular weights are in the thousands or millions such as proteins, DNA, and RNA, in contrast to smaller molecules, e.g., water, salt, etc.

Methylation The process of adding methyl groups (CH_3) at specific sites within a molecule.

Microsome Ribosome plus attached membranous material.

Mitochondrion A substructure within animal cells which is concerned with supplying the energy requirements of the cell, i.e., it is the "power plant" of the cell. Adenosine triphosphate (ATP) is synthesized and stored here for energy requirements.

Neuron A nerve cell—the unit of the nervous system.

Newt A small salamander, an amphibian.

Nucleic acids DNA (deoxyribonucleic acid) and RNA (ribonucleic acid).

Nucleoprotein Protein which is confined to the nucleus of a cell, e.g., histones.

Nucleus A structure within the cell, usually centrally located, which contains DNA and other substances.

Peptide Two amino acids joined together.

Phage A virus which infests a bacterium; also called a bacteriophage.

Phosphorylation Addition of phosphate into specific parts of a molecule.

Physiological saline A solution of sodium chloride (salt) which is similar to that found within the living organism.

Plastid Cell substructure in plants concerned with photosynthesis.

Polypeptide A number of peptides. The difference between a polypeptide and a protein is one of degree; the latter contain more peptides than the former.

Protein A macromolecule which contains many amino acids.

Radioactive labelling The use of radioisotopes to label specific molecules for qualitative and quantitative purposes. For example, one might use thymidine (a DNA nucleotide) containing H^3 at specific sites to determine if any of these labelled nucleotides are present in brain DNA. The presence of labelled DNA would indicate that some DNA synthesis had occurred.

Radioactive precursor A constituent of a molecule containing a radioactive component. For example, uridine-5-H^3 is a radioactive precursor which is incorporated into RNA as it is synthesized.

Recessive gene A gene (the unit of heredity) whose effect does not appear in the organism unless both parents transmit the recessive potential to the offspring, as contrasted with a dominant gene whose effect will appear if the dominant potential is transmitted by one parent.

Reticular formation A diffuse system of nerve cells and tracts in the brain stem which appears to function to produce arousal in higher brain centers. Some individuals maintain that part of this system in the upper brain stem integrates the contributions of lower and higher brain centers to effect organized behavior.

Sedimentation coefficient (Svedberg constant) A number referring to the fact that substances of varying molecular weights reach the bottom of a tube at different rates of centrifugation. Those with high molecular weights (e.g., ribosomal RNA—28S) settle to the bottom before those with low molecular weights (e.g., transfer RNA—5S).

Sucrose gradient The sugar, sucrose, is used to facilitate the separation of different cell components during centrifugation and to provide a pattern of these components according to molecular weights as indicated by sedimentation coefficients (S).

Synapse Point of contact between two neurons. There is no cellular continuity at the synapse but it appears that certain chemicals facilitate communication of one neuron with another neuron at the synapse. For example, in some brain sites acetylcholine (ACh) is synthesized in the presence of the enzyme, choline acetylase (ChA), and stored in vesicles in the presynaptic area. During stimulation ACh is released into the synapse and is assumed to act on the post-synaptic membrane to facilitate the passage of a nerve impulse in the post-synaptic neuron. The enzyme, acetylcholinesterase (AChE), is assumed to attack and degrade ACh. Thus the levels of ChA, ACh, and AChE regulate neural communication in the synaptic area.

Template A molecule which is used as a model for the synthesis of other molecules. For example, DNA acts as a template so that other DNA can be produced with a base sequence complementary to the template DNA, either in the organism or in a test tube; DNA acts as a template so that RNA is synthesized whose base sequence is complementary to that of the template DNA.

Turnover Referring to the synthesis and degradation of molecules.

Vestibular nucleus A group of cells in the lower part of the brain stem (medulla) which contribute to the equilibrium or balance sense.

Virus The simplest form of life. A virus contains a nucleic acid, either DNA or RNA but not both, encased within a protein coat or shell and is capable of self reproduction only within a living cell.

X-ray diffraction A method of determining the molecular structure within substances, usually crystalline substances or those having a highly ordered structure. The X-rays are scattered by electrons in the atomic nuclei of the sample to provide a pattern of the molecular structure.

Psychological—Behavioral

Avoidance conditioning (active) A learning procedure in which a conditioned stimulus (e.g., light) is paired with an unconditioned stimulus (electric shock) such that after a number of trials the unconditioned response (running into a nonshock chamber) occurs in response to the conditioned stimulus. There are two types of active avoidance conditioning procedures:
(a) One-way avoidance—shock is available in only one chamber, and the animal must be placed back in the shock chamber for the next trial after running into nonshock chamber;
(b) two-way avoidance—shock is available in both chambers but for each trial shock is used in only one chamber. Handling of the animal is not required in two-way avoidance, but usually more trials are needed to obtain conditioning with this procedure because the animal must avoid shock by running into a chamber in which he has received a shock on the previous trial.

Conditioned stimulus A stimulus which by association with an unconditioned stimulus acquires the ability to produce the response originally elicited by the unconditional stimulus. For example, in a shock avoidance paradigm, a light a few seconds before the onset of a shock elicits running in a rat after a few trials in which light and shock have been paired.

Escape conditioning A procedure in which an organism learns to use specific behaviors such as running or jumping to a ledge to escape from an electric shock.

Habituation Adaptation to a stimulus after it has repeatedly elicited a specific response, i.e., the response gradually disappears or is reduced in intensity.

Hebb-Williams maze A maze used to test animal intelligence. It consists of a series of 12 related tasks.

Lashley mazes There are three Lashley mazes; all are multiple-choice mazes. They are graded in difficulty with Maze I being the easiest and Maze III the most difficult. These mazes were used by Lashley to evaluate the effects of brain lesions on learning in the rat.

Passive avoidance conditioning A procedure in which an organism must learn to inhibit a response in order to avoid some negative event. For example, a rat learns to press a lever to receive a food pellet. Later when he presses the lever he receives a shock. After a few trials the rat refrains from pressing the lever (even though a food pellet results) so as to avoid electric shock. This conditioning is in contrast to active avoidance conditioning which requires the organism to make a response to avoid a negative event.

Position discrimination learning A task in which the subject chooses one of two, or more, alleyways in which to enter. The presence or absence of reinforcement (e.g., food), or stimulus cues, guides the subject in this choice.

Reversal learning A learning procedure in which an animal learns one task and then the reversal of the task; e.g., in a T maze, learn to turn left and later, to turn right.

Shuttle box A box containing two chambers in which an animal must move from one chamber to the other in order to avoid or escape electrical shock.

T maze A structure in the shape of a T which is used to investigate conditioning. In some cases electric grids are included in the floor of the maze to allow for shocking the organism used. Shock increases the speed of learning to choose a specific arm of the maze.

Unconditioned stimulus A stimulus capable of producing a specific response without training. For example, a shock produces running and escape behavior in a rat.

Water maze A structure containing water used with animals (usually rats or mice) to investigate conditioning. Usually the maze consists of one T unit or a combination of a number of T maze units joined together in a complex pattern.

Y maze A structure in the shape of a Y which is used to investigate conditioning. In some cases electric grids may be included throughout the floors of the maze to provide for shock so as to increase the speed of choosing one of the specific arms by an organism.

REFERENCES

Adair, L. B., Wilson, J. E., and Glassman, E. Brain function and macromolecules, IV. Uridine incorporation into polysomes of mouse brain during different behavioral experiences. *Proc. Natl. Acad. Sci.*, 1968, *61*, 917–22.

Adair, L. B., Wilson, J. E., Zemp, J. W., and Glassman, E. Brain function and macromolecules, III. Uridine incorporation into polysomes of mouse brain during short-term avoidance conditioning. *Proc. Natl. Acad. Sci.*, 1968, *61*, 606–13.

Agranoff, B. W., and Davis, R. E. Evidence for stages in the development of memory. In F. D. Carlson (Ed.), *Physiological and biochemical aspects of nervous integration.* Englewood Cliffs, N. J.: Prentice-Hall, 1968.

Agranoff, B. W., Davis, R. E., and Brink, J. J. Chemical studies on memory fixation in goldfish. *Brain Res.*, 1966, *1*, 303–9.

Agranoff, B. W., Davis, R. E., Casola, L., and Lim, R. Actinomycin D blocks formation of memory of shock avoidance in goldfish. *Science*, 1967, *158*, 1600–1601.

Agranoff, B. W., and Klinger, P. D. Puromycin effect on memory fixation in the goldfish. *Science*, 1964, *146*, 952–53.

Albert, D. J. Memory in mammals: evidence for a system involving nuclear ribonucleic acid. *Neuropsychologia*, 1966, *4*, 79–92.

Allfrey, V. G., and Mirsky, A. E. Mechanisms of synthesis and control of protein and ribonucleic acid synthesis in the cell nucleus. *Cold Spring Harbor Symposium on Quant. Biol.*, 1963, *27*, 247–63.

———. Role of histone in nuclear function. In J. Bonner and P. Ts'O (Eds.), *The nucleohistones.* San Francisco: Holden-Day, 1964.

Allfrey, V. G., Pogo, B. G. T., Pogo, A. O., Kleinsmith, L. J., and Mirsky, A.E . The metabolic behavior of chromatin. In A. V. S. De Reuck and J. Knight (Eds.), *Histones, their role in the transfer of genetic information.* London: J. and A. Churchill, 1966.

Appel, S. A. Effect of inhibition of RNA synthesis on neural information storage. *Nature*, 1965, *207*, 1163–66.

Appel, S. H., Davis, W., and Scott, S. Brain polysomes: response to environmental stimulation. *Science*, 1967, *157*, 836–38.

Appelman, M. M., and Kemp, R. G. Puromycin: a potent metabolic effect independent of protein synthesis. *Biochem. Biophys. Res. Comm.*, 1966, *24*, 564–68.

Applewhite, P. B., and Gardner, F. T. RNA changes during protozoan habituation. *Nature*, 1968, *220*, 1136–37.

Avis H. H., and Carlton, P. L. Retrograde amnesia produced by hippocampal spreading depression. *Science*, 1968, *161*, 73–75.

Babich, F. R., Jacobson, A. L., Bubash, S., and Jacobson, A. Transfer of a response to naïve rats by injection of ribonucleic acid extracted from trained rats. *Science*, 1965, *149*, 656–57.

Barondes, S. H., and Cohen, H. D. Further studies of learning and memory after intracerebral actinomycin-D. *J. Neurochem.*, 1966, *13*, 207–11. (a)

————. Puromycin effect on successive phases of memory storage. *Science*, 1966, *151*, 594–95. (b)

————. Comparative effects of cycloheximide and puromycin on cerebral protein synthesis and consolidation of memory in mice. *Brain Res.*, 1967, *4*, 44–57.

————. Memory impairment after subcutaneous injection of acetoxycycloheximide. *Science*, 1968, *160*, 556–57. (a)

————. Arousal and conversion of "short-term" to "long-term" memory. *Proc. Natl. Acad. Sci.*, 1968, *61*, 923–29. (b)

Barondes, S. H., Dingman, W., and Sporn, M. B. In vitro stimulation of protein synthesis by liver nuclear RNA. *Nature*, 1962, *196*, 145–47.

Barondes, S. H., and Jarvik, M. E. The influences of actinomycin-D on brain RNA synthesis and on memory. *J. Neurochem.*, 1964, *11*, 187–95.

Batkin, S., Woodward, W. T., Cole, R. E., and Hall, J. B. RNA and actinomycin-D enhancement of learning in the carp. *Psychon. Sci.*, 1966, *5*, 345–46.

Beach, G., Emmens, M., Kimble, D. P., and Lickey, M. Autoradiographic demonstration of biochemical changes in the limbic system during avoidance training. *Proc. Natl. Acad. Sci.*, 1969, *62*, 692–96.

Beach, G., and Kimble, D. P. Activity and responsivity in rats after magnesium pemoline injections. *Science*, 1967, *155*, 698–701.

Beerman, W., and Clever, U. Chromosome puffs. *Sci. Amer.*, 1964, *210*, 50–58.

Belozersky, A. N., and Spirin, A. S. Chemistry of the nucleic acids of microorganisms. In E. Chargaff and J. N. Davidson (Eds.), *The nucleic acids. Vol. III.* New York: Academic Press, 1960.

Bendana, F. E., and Galston, A. W. Hormone-induced stabilization of soluble RNA in pea-stem tissue. *Science*, 1965, *150*, 69–70.

Bendich, A., Russell, P. J., and Brown, G. B. On the heterogeneity of the deoxyribonucleic acids. *J. Biol. Chem.*, 1953, *203*, 305–18.

Billen, D., and Hnilica, L. S. Inhibition of DNA synthesis by histones. In J. Bonner and P. Ts'O (Eds.), *The nucleohistones.* San Francisco: Holden-Day, 1964.

Birnstiel, M. L., and Flamm, W. G. Intranuclear site of histone synthesis. *Science*, 1964, *145*, 1435–37.

Birnstiel, M. L., Fleissner, E., and Borek, E. Nucleolus: a center of RNA methylation. *Science*, 1963, *142*, 1577–80.

Bonner, J. The next new biology. *Plant Sci. Bull.*, 1965, *11*, 1–8.

———. Molecular biological approaches to the study of memory. In J. Gaito (Ed.), *Macromolecules and behavior.* New York: Appleton-Century-Crofts, 1966.

Bonner, J., Dahmus, M. E., Fambrough, E., Huang, R. C., Marushige, K., and Yuan, D. Y. H. The biology of isolated chromatin. *Science,* 1968, *159,* 47–56.

Bonner, J., and Huang, R. C. Properties of chromosomal nucleohistone. *J. Mol. Biol.,* 1962, *6,* 169–74. (a)

———. Chromosomal control of enzyme synthesis. *Canad. J. Botany,* 1962, *40,* 1487–97. (b)

———. Role of histone in chromosomal RNA synthesis. In J. Bonner and P. Ts'O (Eds.), *The nucleohistones.* San Francisco: Holden-Day, 1964.

———. Histones as specific repressors of chromosomal RNA synthesis. In A.V.S. De Reuck and J. Knight (Eds.), *Histones: their role in the transfer of genetic information.* London: J. and A. Churchill, 1966.

Bonner, J., Huang, R. C. H., and Gilden, R. V. Chromosomally directed protein synthesis. *Proc. Natl. Acad. Sci.,* 1963, *50,* 893–900.

Bowman, R. E., and Strobel, D. A. Brain RNA metabolism in rat during learning. *J. Comp. Physiol. Psych.,* 1969, *67,* 448–56.

Branch, J. C., and Viney, W. An attempt to transfer a position discrimination habit via RNA extracts. *Psychol. Rep.,* 1966, *19,* 923–26.

Bretscher, M. S. How repressor molecules function. *Nature,* 1968, *217,* 509–11.

Briggs, M. H. and Kitto, G. B. The molecular basis of memory and learning. *Psychol. Rev.,* 1962, *69,* 537–41.

Britten, R. J., and Davidson, E. H. Gene regulation for higher cells: a theory. *Science,* 1969, *165,* 349–57.

Britten, R. J., and Kohne, D. E. Repeated sequences in DNA. *Science,* 1968, *161,* 529–40.

Brown, G. B., and Roll, P. M. Biosynthesis of nucleic acids. In E. Chargaff and J. N. Davidson (Eds.), *The nucleic acids. Vol. II.* New York: Academic Press, 1955.

Brush, T. R., Davenport, J. W., and Polidora, V. J. TCAP: negative results in avoidance and water maze learning and retention. *Psychon. Sci.,* 1966, *4,* 183–84.

Burns, J. T., House, R. F., Fensch, F. C., and Miller, J. G. Effects of magnesium pemoline and dextroamphetamine on human learning. *Science,* 1967, *155,* 849–51.

Busch, H. *Histones and other nuclear proteins.* New York: Academic Press, 1965.

Busch, H., Starbuck, W. C., Singh, E. J., and Ro, T. S. Chromosomal proteins. In M. Locke (Ed.), *The role of chromosomes in development.* New York: Academic Press, 1964.

Busch, H., Steele, W. J., Hnilica, L. S., Taylor, C. W., and Mavioglu, H. Biochemistry of histones and the cell cycle. *J. Cellular Comp. Physiol.,* 1963, *62,* 95–110.

Byrne, R., Levin, J. G., Bladen, H. A., and Nirenberg, M. W. The *in vitro* formation of a DNA-ribosome complex. *Proc. Natl. Acad. Sci.,* 1964, *52,* 140–48.

Byrne, W. L., and Samuel, D. Behavioral modification by injection of brain extract prepared from a trained donor. *Science*, 1966, *154*, 418 (abstract).

Byrne, W. L., and 22 others. Memory transfer. *Science*, 1966, *153*, 658.

Cameron, D. E. The process of remembering. *Brit. J. Psychiat.*, 1963, *109*, 325–33.

Cameron, D. E., and Solyom, L. Effects of ribonucleic acid on memory. *Geriatrics*, 1961, *16*, 74–81.

Carbon, J., David, H., and Studier, M. H. Thiobases in *Escherichia coli* transfer RNA: 2-thiocytosine and 5-methylaminomethyl-2-thiouracil. *Science*, 1968, *161*, 1146–47.

Caskey, C. T., Tompkins, R., Scolnick, E., Caryk, T., and Nirenberg, M. Sequential translation of trinucleotide codons for the initiation and termination of protein synthesis. *Science*, 1968, *162*, 135–38.

Chamberlain, T. J., Halick, P., and Gerard, R. W. Fixation of experience in the rat spinal cord. *J. Neurophysiol.*, 1963, *26*, 663–73.

Chamberlain, T. J., Rothschild, G. H., and Gerard, R. W. Drugs affecting RNA and learning. *Proc. Natl. Acad. Sci.*, 1963, *49*, 918–24.

Chipchase, M. I. H., and Birnstiel, M. L. On the nature of nucleolar RNA. *Proc. Natl. Acad. Sci.*, 1963, *50*, 1101–07.

Clever, U. Puffing in giant chromosomes of diptera and the mechanisms of its control. In J. Bonner and P. Ts'O (Eds.), *The nucleohistones*. San Francisco: Holden-Day, 1964.

Cohen, H. D., and Barondes, S. H. Relationship of degree of training to the effect of acetoxycycloheximide on memory in mice. *75th Annual Conv. APA.*, 1967, *2*, 79–80. (a)

———. Puromycin effect on memory may be due to occult seizures. *Science*, 1967, *157*, 333–34. (b)

Cohen, H. D., Ervin, F., and Barondes, S. H. Puromycin and cycloheximide: different effects on hippocampal electrical activity. *Science*, 1966, *154*, 1557–58.

Cohen, M. J., and Jacklet, J. W. Neurons of insects: RNA changes during injury and regeneration. *Science*, 1965, *148*, 1237–39.

Comb, D. G., Brown, R., and Katz, S. The nuclear DNA and RNA components of the aquatic fungus, *Blastocladiella emersonii*. *J. Mol. Biol.*, 1964, *8*, 781–89.

Comb, D. G., and Katz, S. Studies on the biosynthesis and methylation of transfer RNA. *J. Mol. Biol.*, 1964, *8*, 790–800.

Cook, L., Davidson, A. B., Davis, D. J., Green, H., and Fellows, E. J. Ribonucleic acid: effect on conditioned behavior in rats. *Science*, 1963, *141*, 268–69.

Corning, W. C., and John, E. R. Effect of ribonuclease on retention of conditioned response in regenerated planarians. *Science*, 1961, *134*, 1363–65.

Corson, J. A., and Enesco, H. E. Some effects of injections of ribonucleic acid. *Psychon. Sci.*, 1966, *5*, 217–18.

Crick, F. H. C. The structure of the hereditary material. *Scient. Amer.*, 1954, *191*, 54–61.

———. Nucleic acids. *Scient. Amer.*, 1957, *197*, 188–200.

———. The genetic code. *Scient. Amer.*, 1962, *207*, 66–74.

———. On the genetic code. *Science*, 1963, *139*, 461–64.

Crick, F. H. C., Barnett, L., Brenner, S., and Watts-Tobin, R. J. General nature of the genetic code for proteins. *Nature*, 1961, *192*, 1227–32.

Datta, R. K., and Ghosh, J. J. Studies on the stability of brain cortex ribosomes. *J. Neurochem.*, 1964, *11*, 595–601.

Davis, R. E., and Agranoff, B. W. Stages of memory formation in goldfish: evidence for an environmental trigger. *Proc. Natl. Acad. Sci.*, 1966, *55*, 555–59.

Dellweg, H., Gerner, R., and Wacker, A. Quantitative and qualitative changes in ribonucleic acids of rat brain dependent on age and training experiments. *J. Neurochem.*, 1968, *15*, 1109–19.

Dingman, W., and Sporn, M. B. The incorporation of 8-azaguanine into rat brain RNA and its effect on maze-learning by the rat: an inquiry into the biochemical bases of memory. *J. Psychiat. Res.*, 1961, *1*, 1–11.

———. The isolation and physical characterization of nuclear and microsomal ribonucleic acid from rat brain and liver. *Bioch. Biophy. Acta.*, 1962, *61*, 164–77.

———. Studies on chromatin. I. Isolation and characterization of nuclear complexes of deoxyribonucleic acid, ribonucleic acid, and protein from embryonic and adult tissues of the chicken. *J. Biol. Chem.*, 1964, *239*, 3483–92.

Dyal, J. A., Golub, A. M., and Marrone, R. L. Transfer effects of intraperitoneal injection of brain homogenates. *Nature*, 1967, *214*, 720–21.

Eck, R. V. Genetic code: emergence of a symmetrical pattern. *Science*, 1963, *140*, 477–81.

Edstrom, A. Effect of spinal cord transection on the base composition and content of RNA in the Mauthner nerve fiber of the goldfish. *J. Neurochem.*, 1964, *11*, 557–59.

Edstrom, J. E., and Grampp, W. Nervous activity and metabolism of ribonucleic acids in the crustacean stretch receptor neuron. *J. Neurochem.*, 1965, *12*, 735–41.

Egyhazi, E., and Hydén, H. Experimentally induced changes in the base composition of the ribonucleic acids of isolated nerve cells and their oligodendroglial cells. *J. Biophy. Biochem. Cytol.*, 1961, *10*, 403–10.

Eist, H., and Seal, U. S. The permeability of the blood-brain barrier and blood-CSF barrier to C^{14} tagged ribonucleic acid. *Amer. J. Psychiatry*, 1965, *122*, 584–86.

Enesco, H. E. Fate of ^{14}C-RNA injected into mice. *Exper. Cell Res.*, 1966, *42*, 640–45.

Essman, W. B. Effect of tricyanoaminopropene on the amnesic effect of electroconvulsive shock. *Psychopharmacologia*, 1966, *9*, 426–33.

Essman, W. B., and Lehrer, G. M. Is there a chemical transfer of learning? *Fed. Proc.*, 1966, *25*, 208 (abstract).

Faiszt, J., and Adam, G. Role of different RNA fractions from the brain in transfer effect. *Nature*, 1968, *220*, 367–68.

Fellner, P., and Sanger, F. Sequence analysis of specific areas of the 16S and 23S ribosomal RNAs. *Nature*, 1968, *219*, 236–38.

Filner, P., Wray, J. L., and Varner, J. E. Enzyme induction in higher plants. *Science*, 1969, *165*, 358–67.

Fjerdingstad, E. J., Nissen, T., and Roigaard-Peterson, H. H. Effects of ribonucleic acid (RNA) extracted from the brain of trained animals on learning in rats. *Scand. J. Psych.*, 1965, *6*, 1–6.

———. Facilitation of learning in rats by intracisternal injection of RNA from the brain of trained rats. In O. Walaas (Ed.), *Molecular basis of some aspects of mental activity.* New York: Academic Press, 1966.

Flexner, J. B., and Flexner, L. B. Restoration of expression of memory lost after treatment with puromycin. *Proc. Natl. Acad. Sci.*, 1967, *57*, 1651–54.

Flexner, J. B., Flexner, L. B., and Stellar, E. Memory in mice as affected by intracerebral puromycin. *Science*, 1963, *141*, 57–59.

Flexner, L. B. Dissection of memory in mice with antibiotics. *Proc. Amer. Philos. Soc.*, 1967, *111*, 343–46.

Flexner, L. B., and Flexner, J. B. Effects of acetoxycycloheximide and of an acetoxycycloheximide-puromycin mixture on cerebral protein synthesis and on memory in mice. *Proc. Natl. Acad. Sci.*, 1966, *55*, 369–74.

———. Intracerebral saline: effect on memory of trained mice treated with puromycin. *Science*, 1968, *159*, 330–31.

Flexner, L. B., Flexner, J. B., de la Haba, G., and Roberts, R. B. Loss of memory as related to inhibition of cerebral protein synthesis. *J. Neurochem.*, 1965, *12*, 535–41.

Flexner, L. B., Flexner, J. B., Roberts, R. B., and de la Haba, G. Loss of memory as related to inhibition of cerebral protein synthesis. *Proc. Natl. Acad. Sci.*, 1964, *52*, 1165–69.

Frenster, J. H., Allfrey, V. G., and Mirsky, A. E. Repressed and active chromatin isolated from interphase lymphocytes. *Proc. Natl. Acad. Sci.*, 1963, *50*, 1026–32.

Frey, P. W., and Polidora, V. J. Magnesium pemoline: effect on avoidance conditioning in rats. *Science*, 1967, *155*, 1281–82.

Gaito, J. A biochemical approach to learning and memory. *Psychol. Rev.*, 1961, *68*, 288–92.

———. *Molecular psychobiology: a chemical approach to learning and other behavior.* Springfield, Ill.: C. C. Thomas, 1966.

———. Macromolecules and learning. In G. H. Bourne (Ed.), *The structure and function of nervous tissue. Vol. II. Structure II and physiology.* New York: Academic Press, 1969.

Gaito, J., Davison, J. H., and Mottin, J. Effects of magnesium pemoline on shock avoidance conditioning and on various chemical measures. *Psychon. Sci.*, 1968, *13*, 257–58.

Gaito, J., Mottin, J., and Davison, J. H. Chemical variation in the ventral hippocampus and other brain sites during conditioned avoidance. *Psychon. Sci.*, 1968, *13*, 259–60.

Galand, P., Remy, J., and Ledoux, L. Uptake of exogenous ribonucleic acid by ascites tumor cells. I. Autoradiographic and chromatographic studies. *Exp. Cell Res.*, 1966, *43*, 381–90.

Gambetti, P., Gonatas, N. K., and Flexner, L. B. Puromycin: action on neuronal mitochondria. *Science*, 1968, *161*, 900–902.

Geiger, A. Chemical changes accompanying activity in the brain. In D. Richter (Ed.), *Metabolism of the nervous system*. London: Pergamon, 1957.

Gelfand, S., Clark, L. D., Herbert, E. W., Gelfand, E. D., and Holmes, E. D. Magnesium pemoline: stimulant effects on performance of fatigued subjects. *Clin. Pharmacology Therapeutics*, 1968, *9*, 56–60.

Gerard, R. W. What is memory? *Scient. Amer.*, 1953, *189*, 118–26.

————. The material basis of memory. *J. verbal learning verbal behavior*, 1963, *2*, 22–33.

Gibor, A., and Granick, S. Plastids and mitochondria: inheritable systems. *Science*, 1964, *145*, 890–97.

Gillespie, D., and Spiegelman, S. A quantitative assay for DNA-RNA hybrids with DNA immobilized on a membrane. *J. Mol. Biol.*, 1965, *12*, 829–42.

Glasky, A. J., and Simon, L. N. Magnesium pemoline: enhancement of brain RNA polymerases. *Science*, 1966, *151*, 702–3.

Goldberg, A. L. Memory mechanisms. *Science*, 1964, *144*, 1529.

Goldberg, I. H., Rabinowitz, M., and Reich, E. Basis of actinomycin action. I. DNA binding and inhibition of RNA-polymerase synthetic reactions by actinomycin. *Proc. Natl. Acad. Sci.*, 1962, *48*, 2094–2101.

Goldsmith, L. J. Effect of intra-cerebral actinomycin-D and of electroconvulsive shock on passive avoidance. *J. Comp. Physiol. Psychol.*, 1967, *63*, 126–32.

Gomatos, P. J., and Tamm, J. The secondary structure of reovirus RNA. *Proc. Natl. Acad. Sci.*, 1963, *49*, 707–14.

Goodman, H. M., and Rich, A. Formation of a DNA-soluble RNA hybrid and its relation to the origin, evolution, and degeneracy of soluble RNA. *Proc. Natl. Acad. Sci.*, 1962, *48*, 2101–9.

Goodwin, B. C., and Sizer, I. W. Histone regulation of lactic dehydrogenase in embryonic chick brain tissue. *Science*, 1965, *148*, 242–44.

Gordon, M. W., Deanin, G. G., Leonhordt, H. L., and Gwynn, R. H. RNA and memory: a negative experiment. *Amer. J. Psychiatry*, 1966, *122*, 1174–78.

Gordon, P. Concerning the development of a rational chemotherapy for aging. *Postgraduate Med.*, in press.

Grampp, W., and Edstrom, J. E. The effect of nervous activity on ribonucleic acid of the crustacean receptor neuron. *J. Neurochem.*, 1963, *10*, 725–31.

Granick, S. Cytoplasmic units of inheritance. *Science*, 1965, *147*, 911–13.

Griffith, J. S., and Mahler, H. R. DNA ticketing theory of memory. *Nature*, 1969, *223*, 580–82.

Gross, C. G., and Carey, F. M. Transfer of learned response by RNA injections: failure in attempts to replicate. *Science*, 1965, *150*, 1749.

Grumbach, M. M., Morishima, A., and Taylor, J. H. Human sex chromosome abnormalities in relation to DNA replication and heterochromatinization. *Proc. Natl. Acad. Sci.*, 1963, *49*, 581–89.

Grunberg-Manago, M. Enzymatic synthesis of nucleic acids. In *Progress in Biophysics, Vol. 13*. New York: Pergamon Press, 1963.

Gurowitz, E. M. Some effects of injections of brain homogenates on behavior. *Psychol. Reports*, 1967, *23*, 899–910.

————. *The molecular basis of memory.* Englewood Cliffs, N.J.: Prentice-Hall, 1969.

Gurowitz, E. M., Gross, D. A., and George, R. Effects of TCAP on passive avoidance learning in the rat. *Psychon. Sci.,* 1968, *12,* 293–94.

Gurowitz, E. M., Lubar, J. P., Ain, B. R., and Gross, D. G. Disruption of passive avoidance learning by magnesium pemoline. *Psychon. Sci.,* 1967, *8,* 19–20.

Guthrie, C., and Nomura, M. Initiation of protein synthesis: a critical test of the 30S subunit model. *Nature,* 1968, *219,* 232–35.

Gutierez, R. M., and Hnilica, L. S. Tissue specificity of histone phosphorylation. *Science,* 1967, *157,* 1324–25.

Halas, E. S. Bradfield, K., Sandlie, M. E., Theye, F., and Beardsley, J. Changes in rat behavior due to RNA injection. *Physiology and Behavior,* 1966, *1,* 281–83.

Halstead, W. C. Brain and Intelligence. In L. A. Jeffres (Ed.), *Cerebral mechanisms in behavior.* New York: Wiley, 1951.

Hamilton, T. H. Control by estrogen of genetic transcription and translation. *Science,* 1968, *161,* 649–60.

Harbers, E., Domagk, G. F., and Muller, W. *Introduction to nucleic acids.* New York: Reinhold, 1968.

Hardesty, B., Miller, R., and Schwett, R. Polyribosome breakdown and hemoglobin synthesis. *Proc. Natl. Acad. Sci.,* 1963, *50,* 924–31.

Hayashi, M., Hayashi, M. N. and Spiegelman, S. Replicating form of a single-stranded DNA virus: isolation and properties. *Science,* 1963, *140,* 1313–16.

Hayashi, M., Spiegelman, S., Franklin, N. C., and Luria, S. E. Separation of the RNA message transcribed in response to a specific inducer. *Proc. Natl. Acad. Sci.,* 1963, *49,* 729–36.

Hebb, D. O. *The organization of behavior.* New York: Wiley, 1949.

Hechter, O., and Halkerston, J. D. K. Effects of steroid hormones on gene regulation and cell metabolism. In V. E. Hall, A. C. Grese, and R. R. Sonnerschein (Eds.), *Annual review of physiology. Vol. XXVII.* Palo Alto: Annual Reviews, 1965.

Heinbecker, P., and Bartley, S. H. Manner of strychnine action on nervous system. *Amer. J. Physiol.,* 1939, *125,* 172–87.

Henney, H. R. Jr., and Storck, R. Polyribosomes and morphology in *Neurospora crassa. Proc. Natl. Acad. Sci.,* 1964, *51,* 1050–55.

Herbert, E. W., Gelfand, S., Clark, L. D., and Gelfand, D. M. Magnesium pemoline: stimulant effects on performance of fatigued men. *Clin. Pharmacology Therapeutics,* 1968, *9,* 578–81.

Hnilica, L. S., and Bess, L. G. The heterogeneity of arginine-rich histones. *Anal. Biochem.,* 1965, *12,* 421–36.

Hoagland, M. B., Scornik, O. A., and Pfefferkorn, L. C. Aspects of control of protein synthesis in normal and regenerating rat liver. II. A microsomal inhibitor of amino acid incorporation whose action is antagonized by guanosine triphosophate. *Proc. Natl. Acad. Sci.,* 1964, *51,* 1184–91.

Holley, R. W., Apgar, J., Everett, G. A., Madison, J. T., Marquisse, M., Merrill, S. H., Penswick, J. R., and Zamir, A. Structure of a ribonucleic acid. *Science,* 1965, *147,* 1462–65.

Huang, R. C., and Bonner, J. Histone, a suppressor of chromosomal RNA synthesis. *Proc. Natl. Acad. Sci.*, 1962, *48*, 1216–22.

Hurwitz, J., and August, J. T. The role of DNA in RNA synthesis. In J. N. Davidson and W. E. Cohn (Eds.), *Progress in nucleic acid research.* New York: Academic Press, 1963.

Hurwitz, J., and Furth, J. J. Messenger RNA. *Scientific Amer.*, 1962, *206*, 41–49.

Hydén, H. Biochemical changes in glial cells and nerve cells at varying activity. In *Proc. 4th. Intern. Congr. Biochem. Biochemistry of the central nervous system. Vol. III.* London: Pergamon Press, 1959.

————. Satellite cells in the nervous system. *Scientific Amer.*, 1961, *205*, 62–70.

————. RNA in brain cells. In G. C. Quarton, T. Melnechuk, and F. O. Schmitt (Eds.), *The neurosciences.* New York: The Rockefeller Univ. Press, 1967.

Hydén, H., and Egyhazi, E. Nuclear RNA changes of nerve cells during a learning experiment in rats. *Proc. Natl. Acad. Sci.*, 1962, *48*, 1366–73.

————. Glial RNA changes during a learning experiment with rats. *Proc. Natl. Acad. Sci.*, 1963, *49*, 618–24.

————. Changes in RNA and base composition in cortical neurons of rats in a learning experiment involving transfer of handedness. *Proc. Natl. Acad. Sci.*, 1964, *52*, 1030–35.

Hydén, H., Egyhazi, E., John, E. R., and Bartlett, F. RNA base ratio changes in planaria during conditioning. *J. Neurochem.*, 1969, *16*, 813–21.

Hydén, H., and Lange, P. W. A differentiation in RNA response in neurons early and late during learning. *Proc. Natl. Acad. Sci.*, 1965, *53*, 946–52.

————. A genic stimulation with production of adenine-uracil rich RNA in neurons and glia in learning. The question of transfer of RNA from glia to neurons. *Die Naturwissenschaften*, 1966, *53*, 64–70.

Izawa, M., Allfrey, V. G., and Mirsky, A. E. The relationship between RNA synthesis and loop structure in lampbrush chromosomes. *Proc. Natl. Acad. Sci.*, 1963, *49*, 544–51.

Jacob, F., and Monod, J. Genetic regulatory mechanisms in the synthesis of protein. *J. Mol. Biol.*, 1961, *3*, 318–56.

Jacob, J., and Sirlin, J. L. Synthesis of RNA *in vitro* stimulated in Dipterian salivary glands by 1, 1, 3-Tricyano-2-amino-2-amino-1-propene. *Science*, 1964, *144*, 1011–12.

Jacobson, A. L., Babich, F. R., Bubash, S., and Goren, C. Maze preferences in naive rats produced by injection of ribonucleic acid from trained rats. *Psychonomic Science*, 1966, *4*, 3–4.

John, E. R. *Mechanisms of memory.* New York: Academic Press, 1967.

Jones, C. T., and Banks, P. Inhibition of respiration by puromycin in slices of cerebral cortex. *J. Neurochem.*, 1969, *16*, 825–28.

Jukes, T. H. Observations on the possible nature of the genetic code. *Biochem. Biophys. Res. Comm.*, 1963, *10*, 155–59.

Karlson, P. Chemical and immunological aspects of hormones on the chemistry and mode of action in insect hormones. *Gen. Comp. Endocrinology*, 1962, *S1*, 1–7.

Karol, M. H., and Simpson, M. V. DNA biosynthesis by isolated mitochondria: a replicative rather than a repair process. *Science*, 1968, *162*, 470–73.

Katz, J. J., and Halstead, W. C. Protein organization and mental function. *Comp. Psychol. Monogr.*, 1950, *20*, No. 103, 1–38.

Kenney, F. T., and Kull, F. J. Hydrocortisone-stimulated synthesis of nuclear RNA in enzyme induction. *Proc. Natl. Acad. Sci.*, 1963, *50*, 493–99.

Kondo, M., Eggerston, G., Eisenstadt, J., and Lengyel, P. Ribosome formation from subunits: dependence on formylmethionyl-transfer RNA in extracts from *E. coli. Nature*, 1968, *220*, 368–70.

Kornberg, A. The synthesis of DNA. *Scientific Amer.*, 1968, *219*, 64–78.

Kral, V. A., and Sved, S. Clinical and biochemical remarks on the ribonucleic acid treatment of Alzheimer's disease. Paper presented in symposium entitled "Nucleic Acids and Behavior," Midwestern Psychological Association meetings, Chicago, Illinois, May 2, 1963.

Landauer, T. K., and Eldridge, L. Failure of actinomycin-D to inhibit passive avoidance learning: a confirmation. *74th Ann. Proc. Amer. Psych. Assn.*, 1966, *2*, 123–24.

Landgridge, R., and Gomatos, P. J. The structure of RNA. *Science*, 1963, *141*, 694–98.

Leslie, I. The nucleic acid content of tissues and cells. In E. Chargaff and J. N. Davidson (Eds.), *The nucleic acids: chemistry and biology, Vol. I*. New York: Academic Press, 1955.

Littau, V. C., Allfrey, V. G., Frenster, J. H., and Mirsky, A. E. Active and inactive regions of nuclear chromatin as revealed by electron microscope autoradiography. *Proc. Natl. Acad. Sci.*, 1964, *52*, 93–100.

Loewy, A. G., and Siekevitz, P. *Cell structure and function*. New York: Holt, Rinehart and Winston, 1963.

Lubar, J. F., Boitano, J. J., Gurowitz, E. M., and Ain, B. R. Enhancement of performance in the Hebb-Williams maze by magnesium pemoline. *Psychon. Sci.*, 1967, *7*, 381–82.

Luttges, M., Johnson, T., Buck, C., Holland, J., and McGaugh, J. An examination of "transfer of learning" by nucleic acid. *Science*, 1966, *151*, 834–37.

Machlus, B., and Gaito, J. Detection of RNA species unique to a behavioral task. *Psychon. Sci.*, 1968, *10*, 253–54. (a)

———. Unique RNA species developed during a shock avoidance task. *Psychon. Sci.*, 1968, *12*, 111–12. (b)

———. The use of successive competition hybridization to detect RNA species in a shock avoidance task. *Nature*, 1969, *222*, 573–74.

Mandel, P., Harth, S., and Borkowski, T. Metabolism of the nucleic acids in various zones of the brain. In S. S. Kety and J. Elkes (Eds.), *Regional neurochemistry*. New York: Pergamon Press, 1961.

Markert, C. L. Developmental genetics. In *The Harvey Lectures, Series 59*. New York: Academic Press, 1965.

Markert, C. L., and Ursprung, H. Production of replicable persistent changes in zygote chromosomes of *Rana pipiens* by injected proteins from adult liver nuclei. *Dev. Biol.*, 1963, *7*, 560–77.

Marushige, K., and Bonner, J. Template properties of liver chromatin. *J. mol. Biol.*, 1966, *15*, 160–74.

Matthaei, J. H., Jones, O. W., Martin, R. G., and Nirenberg, M. W. Characteristics and composition of RNA coding units. *Proc. Natl. Acad. Sci.*, 1962, *48,* 666–77.

McConnell, J. V. RNA and memory. Paper presented in Symposium on the Role of Macromolecules in Complex Behavior, Kansas State University, Manhattan, 1964.

McGaugh, J. Time-dependent processes in memory storage. *Science,* 1966, *153,* 1351–58.

McGaugh, J. L., and Petrinovich, L. Effects of drugs on learning and memory. *Int. Rev. Neurobiol.,* 1965, *8,* 139–96.

Meyerson, F. Z., Kruglikov, R. L., and Kolomeitsova, I. A. The role of the synthesis of nucleic acids in the mechanism of stabilization of conditioned reflexes and memory. *Bull. Exp. Biol. Med.,* 1965, *12,* 6–7.

Mills, D. R., Peterson, R. L., and Spiegelman, S. An extracellular Darwinian experiment with a self-duplicating nucleic acid molecule. *Proc. Natl. Acad. Sci.,* 1967, *58,* 217–24.

Miyagi, M., Kohl, D., and Flickinger, R. A. Detection of qualitative differences between the RNA of livers and kidneys of adult chickens and a temporal and regional study of liver RNA in chick embryo. *J. Exp. Zoology,* 1967, *165,* 147–53.

Morris, M. R., Aghajanian, G. K., and Bloom, F. E. Magnesium pemoline: failure to affect *in vivo* synthesis of brain RNA. *Science,* 1967, *155,* 1125–26.

Nasello, A. G., and Izquierdo, I. Effect of learning and of drugs on the ribonucleic acid concentration of brain structures of the rat. *Exp. Neurology,* 1969, *23,* 521–28.

Neidle, A., and Waelsch, H. Histones: species and tissue specificity. *Science,* 1964, *145,* 1059–61.

Nirenberg, M. W. The genetic code: II. *Sci. Amer.,* 1963, *208,* 80–94.

Nissen, J. H., Roigaard-Peterson, H. H., and Fjerdingstad, F. J. Effect of ribonucleic acid (RNA) extracted from the brain of trained animals on learning in rats. II. Dependence of RNA effect on training conditions prior to RNA extraction. *Scand. J. Psych.,* 1965, *6,* 265–72.

Ochoa, S. Enzymatic mechanisms in the transmission of genetic information. In M. Kasha and B. Pullman (Eds.), *Horizons in biochemistry.* New York: Academic Press, 1962.

Ohtaka, Y., and Spiegelman, S. Translational control of protein synthesis in a cell-free system directed by a polycistronic viral RNA. *Science,* 1963, *142,* 493–97.

Olson, R. E. Vitamin K induced prothrombin formation: antagonism by actinomycin-D. *Science,* 1964, *145,* 926–28.

O'Malley, B. W., Aronow, A., Peacock, A. C., and Dingman, C. W. Estrogen-dependent increase in transfer RNA during differentiation of the chick oviduct. *Science,* 1968, *162,* 567–68.

Oshima, K., Gorbman, A., and Shimada, H. Memory-blocking agents: effects on olfactory discrimination in homing salmon. *Science,* 1969, *165,* 86–88.

Painter, T. S. Fundamental chromosome structure. *Proc. Natl. Acad. Sci.,* 1964, *51,* 1282–90.

Pardee, A. B. Aspects of genetic and metabolic control of protein synthesis. In J. M. Allen (Ed.), *The molecular control of cellular activity*. New York: McGraw-Hill, 1962.

Penfield, W. Neurophysiological basis of the higher functions of the nervous system —introduction. In J. Field (Ed.), *Handbook of physiology-neurophysiology, Vol. III*. Baltimore: Williams and Wilkins, 1960.

Perry, R. P., Srinivasan, P. R., and Kelley, D. E. Hybridization of rapidly labeled nuclear ribonucleic acids. *Science*, 1964, *145*, 504–7.

Pevzner, L. C. Nucleic acid changes during behavioral events. In J. Gaito (Ed.), *Macromolecules and behavior*, first edition. New York: Appleton-Century-Crofts, 1966.

Plotnikoff, N. Magnesium pemoline: enhancement of learning and memory of a conditioned avoidance response. *Science*, 1966, *151*, 703–4. (a)

———. Magnesium pemoline: enhancement of memory after electroconvulsive shock in rats. *Life Sciences*, 1966, *5*, 1495–98. (b)

———. Pemoline and magnesium hydroxide: memory consolidation following acquisition trials. *Psychon. Sci.*, 1967, *9*, 141–42.

Pogo, B. G. T., Allfrey, V. G., and Mirsky, A. E. RNA synthesis and histone acetylation during the course of gene activation in lymphocytes. *Proc. Natl. Acad. Sci.*, 1966, *55*, 805–12.

Pollard, E. C. Ionizing radiation: effect on genetic transcription. *Science*, 1964, *146*, 927–29.

Ptashne, M. Specific binding of the λ phage repress to λDNA. *Nature*, 1967, *214*, 232–34.

Reid, B. R., and Cole, R. D. Biosynthesis of a lysine-rich histone in isolated calf thymus nuclei. *Proc. Natl. Acad. Sci.*, 1964, *51*, 1044–50.

Reinis, S. The formation of conditioned reflexes after the parental administration of brain homogenate. *Activ. Nerv. Super.*, 1965, *7*, 167–68.

Rich, A. On the problems of evolution and biochemical information transfer. In M. Kasha and B. Pullman (Eds.), *Horizons in biochemistry*. New York: Academic Press, 1962.

Ris, H. The structure of nucleohistones in chromosomes. *Science*, 1964, *146*, 428–29.

Ritossa, F. M., and Spiegelman, S. Localization of DNA complementary to ribosomal RNA in the nucleolus organizer region of *Drosophilia melanogaster*. *Proc. Natl. Acad. Sci.*, 1965, *53*, 737–45.

Roberts, R. B. Further implications of the doublet code. *Proc. Natl. Acad. Sci.*, 1962, *48*, 1245–50.

Rosenblatt, F., Farrow, J. T., and Rhine, S. The transfer of learned behavior from trained to untrained rats by means of brain extracts. Part I. *Proc. Natl. Acad. Sci.*, 1966, *55*, 548–55. (a)

———. The transfer of learned behavior from trained to untrained rats by means of brain extracts. Part II. *Proc. Natl. Acad. Sci.*, 1966, *55*, 787–92. (b)

Sager, R. and Ryan, F. J. *Cell heredity*. New York: Wiley, 1961.

Sampson, M., Katoh, A., Hotta, Y., and Stern, H. Metabolically labile deoxyribonucleic acid. *Proc. Natl. Acad. Sci.*, 1963, *50*, 459–63.

Santer, M. Ribosomal RNA on the surface of ribosomes. *Science,* 1962, *141,* 1049–50.

Schaeffer, E. The effect of RNA injections on shock avoidance conditioning and on brain chemistry. Unpublished M. A. thesis, York University, 1967.

Schneiderman, H. A., and Gilbert, L. J. Control of growth and development in insects. *Science,* 1964, *143,* 325–33.

Schwarz, M. R., and Rieke, W. O. Appearance of radioactivity in mouse cells after administration of labelled macromolecular RNA. *Science,* 1962, *136,* 152–53.

Scott, R. B., and Bell, E. Protein synthesis during development: control through messenger RNA. *Science,* 1964, *145,* 711–14.

Sehon, A. Stereospecificity. *Science,* 1965, *148,* 401–11.

Shashoua, V. E. RNA changes in goldfish during learning. *Nature,* 1968, *217,* 238–40.

————. RNA metabolism in goldfish brain during acquisition of new behavioral patterns. *Proc. Natl. Acad. Sci.,* 1970, *65,* 160–67.

Simon, L. N., and Glasky, A. J. Magnesium pemoline: enhancement of brain RNA synthesis *in vivo. Life Sciences,* 1968, *7,* 197–202.

Singer, M. F. 1968 Nobel laureate in medicine or physiology. *Science,* 1968, *162,* 433–36.

Smellie, R. M. S. The biosynthesis of ribonucleic acid in animal systems. In J. N. Davidson and W. E. Cohn (Eds.), *Progress in nucleic acid research.* Vol. I. New York: Academic Press, 1963.

Smith, C. E. Is memory a matter of enzyme induction? *Science,* 1962, *138,* 889–90.

Smith, R. G. Magnesium pemoline: lack of facilitation in human learning, memory, and performance tests. *Science,* 1967, *155,* 603–5.

Solyom, L., and Gallay, II. M. Effect of malononitrile dimer on operant and classical conditioning of aged white rats. *Internat. J. Neuropsychiat.,* 1966, *2,* 577–84.

Sonneborn, T. M. The differentiation of cells. *Proc. Natl. Acad. Sci.,* 1964, *51,* 915–29.

————. Nucleotide sequence of a gene: first complete specification. *Science,* 1965, *148,* 1410.

Sporn, M. B., and Dingman, C. W. Histone and DNA in isolated nuclei from chicken brain, liver, and erythrocytes. *Science,* 1963, *140,* 316–18.

Srinivasan, P. R., and Borek, E. Enzymatic alteration of nucleic acid structure. *Science,* 1964, *145,* 548–53.

Staehelin, T., Wetlstern, F. O., and Noll, H. Breakdown of rat-liver ergosomes *in vivo* after actinomycin inhibition of messenger RNA synthesis. *Science,* 1963, *140,* 180–83.

Stanley, W. M., and Valens, E. G. *Viruses and the nature of life.* New York: E. P. Dutton, 1961.

Stedman, E., and Stedman, E. Cell specificity of histones. *Nature,* 1950, *166,* 780–81.

Stein, H. H., and Yellin, T. O. Pemoline and magnesium hydroxide: lack of effect on RNA and protein synthesis. *Science,* 1967, *157,* 96–97.

Stevenin, J., Samec, J., Jacob, M., aand Mandel P. Determination de la fraction du genome codant pour les RNA ribosomiques et messagers dans le cerveau du rat adulte. *J. Mol. Biol.*, 1968, *33*, 777–93.

Strauss, B. S. *An outline of chemical genetics.* Philadelphia: Saunders, 1960.

Sved, S. The metabolism of exogenous ribonucleic acid injected into mice. *Canad. J. Biochem.*, 1965, *43*, 949–58.

Swift, H. Nucleic acids and cell morphology in dipterian salivary glands. In J. M. Allen (Ed.), *The molecular control of cellular activity.* New York: McGraw-Hill, 1962.

――――. In J. Bonner and P. Ts'O (Eds.), *The nucleohistones.* San Francisco: Holden-Day, 1964.

Talwar, G. P., Sharma, S. K., and Gupta, S. L. The enigma of the mechanism of action of hormones. *J. Scientific Ind. Res.*, 1968, *27*, 28–29.

Tatum, E. L. Medicine and molecular genetics. *Bull. N.Y. Acad. Sci.*, 1964, *40*, 1–9.

Thompson, R., and McConnell, J. V. Classical conditioning in the planarian, *Dugesia dorotocephola. J. Comp. physiol. Psychol.*, 1955, *48*, 65–68.

Tomkins, G. M., Gelehrter, T. D., Granner, D., Martin, P. Jr., Samuels, H. H., and Thompson, E. B. Control of specific gene expression in higher organisms. *Science*, 1969, *166*, 1474–80.

Ungar, G. Biological assays for the molecular coding of acquired information. In J. Gaito (Ed.), *Macromolecules and brain function*, 2nd ed. New York: Appleton-Century-Crofts, 1971.

Ungar, G., and Cohen, M. Induction of morphine tolerance by material extracted from brain of tolerant animals. *Intern. J. Neuropharmacol.*, 1965, *5*, 1–10.

Ungar, G., and Oceguerra-Navarro, C. Transfer of habituation by material extracted from brain. *Nature*, 1965, *207*, 301–2.

Viney, W., Branch, J. C., and Gill, W. E. Facilitation of discrimination learning by injection of an RNA extract prepared from donor subjects. *Psychol. Rep.*, 1967, *21*, 601–5.

Wagner, A. R., Gardner, J. B., and Beatty, W. W. Yeast ribonucleic acid: effects on learned behavior in the rat. *Psychon. Sci.*, 1966, *4*, 33–34.

Warner, J. R., and Rich, A. The number of soluble RNA molecules on reticulocyte polyribosomes. *Proc. Natl. Acad. Sci.*, 1964, *51*, 1134–41.

Warner, J. R., Rich, A., and Hall, C. E. Electron microscope studies of ribosomal clusters synthesizing hemoglobin. *Science*, 1962, *138*, 1399–1403.

Watson, J. D. *Molecular biology of the gene.* New York: Benjamin, 1965.

Wicks, W. D., and Kenney, F. T. RNA synthesis in rat seminal vesicles: stimulation by testoterone. *Science*, 1964, *144*, 1346–47.

Wilkins, M. H. F. Molecular configuration of nucleic acids. *Science*, 1963, *140*, 941–50.

Wool, I. G., and Munro, A. J. An influence of insulin on the synthesis of a rapidly labeled RNA isolated rat diaphragm. *Proc. Natl. Acad. Sci.*, 1963, *50*, 918–23.

Yamagami, S., Kawakita, Y., and Naka, S. Base composition of RNA of the subcellular fractions from guinea pig brains. *J. Neurochem.*, 1964, *11*, 899–900.

Yankofsky, S. A., and Spiegelman, S. Distinct cistrons for the two ribosomal RNA components. *Proc. Natl. Acad. Sci.*, 1963, *49*, 538–44.

Zelman, A., Kabat, L., Jacobson, R., and McConnell, J. V. Transfer of training through injection of "conditioned" RNA into untrained planarians. *Worm Runner's Digest*, 1963, *5*, No. 1, 14–21.

Zemp, J. W., Wilson, J. E., Schlesinger, K., Boggan, W. O., and Glassman, E. Brain function and macromolecules, I. Incorporation of uridine into RNA of mouse brain during short-term training experiences. *Proc. Natl. Acad. Sci.*, 1966, *55*, 1423–31.

Zemp, J. W., Wilson, J. E., and Glassman, E. Brain function and macromolecules, II. Site of increased labelling of RNA in brains of mice during a short-term training experience. *Proc. Natl. Acad. Sci.*, 1967, *58*, 1120–25.

Zubay, G. Molecular model for protein synthesis. *Science*, 1963, *140*, 1092–95.

———. Nucleohistone structure and function. In J. Bonner aand P. Ts'O (Eds.), *The nucleohistones.* San Francisco: Holden-Day, 1964.

INDEX

Lipids, 67, 68, 89
Liver, 51, 55, 57

Magnesium pemoline, 3, 62, 72–74, 84, 93
Malononitrile (tricyanoaminopropene), 49, 71–72
Memory, 44, 60, 72, 73, 74–81, 83, 86, 87, 89, 96, 97
Memory consolidation, 93, 94
Methylation, 13, 90
Microsome, 27
Mitochondria, 15, 80, 81
Molecular biology, 1, 2, 88
Molecular psychobiology, 1, 2, 43, 92

Nerve cell (neurons), 45, 46, 47, 48, 87, 89, 90
Nerve impulse, 1, 87
Nucleolus, 9
Nucleosides, 5, 9, 12, 68
Nucleotides, 5, 6, 7, 9, 11, 12, 13, 22, 24, 25, 26, 38, 62, 67
Nucleus, 8, 9, 12, 21, 27, 31, 33, 35, 39, 86

Parallel effects, 44, 45, 62, 68, 71, 80, 82, 84
Phenylalanine, 2
Phenylketonuria, 2
Phenylpyruvic acid, 2
Phospholipids, 62
Phosphoprotein, 39, 40
Phosphorylation, 39, 40, 41
Planarian, 49, 63, 69
Polypeptide, 22, 23, 65, 67, 74, 75, 90
Polyribosomes (polysomes), 22, 23, 50, 51, 96, 97
Postural asymmetry, 71–72
Primary effects, 43–44, 45, 60, 62, 68, 70, 80, 82, 84, 91
Protein, 1, 2, 3, 10, 11, 12, 16, 17, 18, 19, 21, 22, 23, 24, 25, 28, 36, 37, 43, 44, 45, 51, 65, 66, 67, 68, 69, 71, 80, 81, 82, 83, 84, 86, 87, 88, 89, 90, 93, 94, 95, 96
Protein synthesis, 1, 10, 11, 13, 14, 17, 21–26, 27, 28, 29, 36, 39, 44, 62, 69, 73, 74–81, 82, 83, 84, 89, 90, 96, 97
Purines, 8, 9, 88
Puromycin, 34, 74–81
Pyrimidine, 8, 9, 88

Recessive gene, 2
Regulatory events, 31–42
Repressor, 16, 17, 28, 29, 30
Reticular formation, 30, 46
Ribonuclease (RNase), 12, 53, 60, 65, 69, 97
Ribonucleic acid (RNA), 1, 2, 3, 5, 8–15, 16, 17, 18, 19, 21, 27, 28, 29, 30, 31, 33, 34, 35, 36, 37, 38, 39, 40, 41, 43–52, 53–59, 60–68, 69–81, 82–84, 87–91, 92, 93, 95, 96, 97
activator, 13, 16, 30, 38, 58
chromosomal, 13, 37, 38
messenger, 9, 10, 11, 14, 15, 16, 21, 22, 23, 24, 25, 26, 28, 29, 30, 34, 35, 36, 48, 58, 76, 80, 89, 90
ribosomal, 9, 10, 11, 13, 14, 15, 34, 48, 50, 63, 89
transfer, 9, 11, 12, 13, 14, 16, 17, 21, 22, 23, 26, 48, 50
yeast, 60–62, 73, 74
Ribonucleic acid (RNA) synthesis, 17, 19, 26, 27, 28, 29, 30, 31, 33, 34, 35, 36, 37, 38, 39, 40, 41, 44, 51, 52, 69, 70, 71, 72, 73, 80, 81, 82, 83, 84, 89, 90, 96, 97
Ribosomes, 10, 11, 17, 21, 22, 23, 34, 35, 36, 80, 89, 90
RNA polymerase, 17, 37, 72, 73
RNA therapy, 3, 60, 94

Secondary effects, 44, 45, 62, 68, 80, 82, 84
Selective model, 87, 88–90, 91
Sickle cell anemia, 2
Strychnine sulfate, 3, 93, 94
Synapse, 86, 89, 90
Synthetic polyribonucleotides, 24, 25

Thymine, 6, 7, 8, 9, 14
Transfer experiment, 62–68, 93
Transmitter substance, 87, 89
Tricyanoaminopropene (malononitrile), 49, 71–72, 74, 84, 93
Trinucleotide coding, 11, 24
Tyrosine, 2

Uracil, 8, 9, 14, 45–50, 53, 89
Uridine, 9, 54, 55

Virus, 3, 7, 14, 15, 28, 92